MANUALS AND REPORTS ON ENGINEERING PRACTICE

(As developed by the ASCE Technical Procedures Committee, July 1930, and revised March 1935, February 1962, and April 1982)

A manual or report in this series consists of an orderly presentation of facts on a particular subject, supplemented by an analysis of limitations and applications of these facts. It contains information useful to the average engineer in his or her everyday work, rather than findings that may be useful only occasionally or rarely. It is not in any sense a "standard," however; nor is it so elementary or so conclusive as to provide a "rule of thumb" for nonengineers.

Furthermore, material in this series, in distinction from a paper (which expresses only one person's observations or opinions), is the work of a committee or group selected to assemble and express information on a specific topic. As often as practicable, the committee is under the direction of one or more of the Technical Divisions and Councils, and the product evolved has been subjected to review by the Executive Committee of the Division or Council. As a step in the process of this review, proposed manuscripts are often brought before the members of the Technical Divisions and Councils for comment, which may serve as the basis for improvement. When published, each work shows the names of the committees by which it was compiled and indicates clearly the several processes through which it has passed in review, in order that its merit may be definitely understood.

In February 1962 (and revised in April 1982) the Board of Direction voted to establish a series entitled "Manuals and Reports on Engineering Practice," to include the Manuals published and authorized to date, future Manuals of Professional Practice, and Reports on Engineering Practice. All such Manual or Report material of the Society would have been refereed in a manner approved by the Board Committee on Publications and would be bound, with applicable discussion, in books similar to past Manuals. Numbering would be consecutive and would be a continuation of present Manual numbers. In some cases of reports of joint committees, bypassing of Journal publications may be authorized.

MANUALS AND REPORTS ON ENGINEERING PRACTICE

No.	Title
13	Filtering Materials for Sewage Treatment Plants
14	Accommodation of Utility Plant Within the Rights-of-Way of Urban Streets and Highways
35	A List of Translations of Foreign Literature on Hydraulics
40	Ground Water Management
41	Plastic Design in Steel: A Guide and Commentary
45	Consulting Engineering: A Guide for the Engagement of Engineering Services
46	Pipeline Route Selection for Rural and Cross-Country Pipelines
47	Selected Abstracts on Structural Applications of Plastics
49	Urban Planning Guide
50	Planning and Design Guidelines for Small Craft Harbors
51	Survey of Current Structural Research
52	Guide for the Design of Steel Transmission Towers
53	Criteria for Maintenance of Multilane Highways
54	Sedimentation Engineering
55	Guide to Employment Conditions for Civil Engineers
57	Management, Operation and Maintenance of Irrigation and Drainage Systems
59	Computer Pricing Practices
60	Gravity Sanitary Sewer Design and Construction, Second Edition
62	Existing Sewer Evaluation and Rehabilitation
63	Structural Plastics Design Manual
64	Manual on Engineering Surveying
65	Construction Cost Control
66	Structural Plastics Selection Manual
67	Wind Tunnel Studies of Buildings and Structures
68	Aeration: A Wastewater Treatment Process
69	Sulfide in Wastewater Collection and Treatment Systems
70	Evapotranspiration and Irrigation Water Requirements
71	Agricultural Salinity Assessment and Management
72	Design of Steel Transmission Pole Structures
73	Quality in the Constructed Project: A Guide for Owners, Designers, and Constructors
74	Guidelines for Electrical Transmission Line Structural Loading
76	Design of Municipal Wastewater Treatment Plants
77	Design and Construction of Urban Stormwater Management Systems
78	Structural Fire Protection
79	Steel Penstocks

No.	Title
80	Ship Channel Design
81	Guidelines for Cloud Seeding to Augment Precipitation
82	Odor Control in Wastewater Treatment Plants
83	Environmental Site Investigation
84	Mechanical Connections in Wood Structures
85	Quality of Ground Water
86	Operation and Maintenance of Ground Water Facilities
87	Urban Runoff Quality Manual
88	Management of Water Treatment Plant Residuals
89	Pipeline Crossings
90	Guide to Structural Optimization
91	Design of Guyed Electrical Transmission Structures
92	Manhole Inspection and Rehabilitation
93	Crane Safety on Construction Sites
94	Inland Navigation: Locks, Dams, and Channels
95	Urban Subsurface Drainage
96	Guide to Improved Earthquake Performance of Electric Power Systems
97	Hydraulic Modeling: Concepts and Practice
98	Conveyance of Residuals from Water and Wastewater Treatment
99	Environmental Site Characterization and Remediation Design Guidance
100	Groundwater Contamination by Organic Pollutants: Analysis and Remediation
101	Underwater Investigations
102	Design Guide for FRP Composite Connections
103	Guide to Hiring and Retaining Great Civil Engineers
104	Recommended Practice for Fiber-Reinforced Polymer Products for Overhead Utility Line Structures
105	Animal Waste Containment in Lagoons
106	Horizontal Auger Boring Projects
107	Ship Channel Design and Operation
108	Pipeline Design for Installation by Horizontal Directional Drilling
109	Biological Nutrient Removal (BNR) Operation in Wastewater Treatment Plants
110	Sedimentation Engineering: Processes, Measurements, Modeling, and Practice
111	Reliability-Based Design of Utility Pole Structures
112	Pipe Bursting Projects
113	Substation Structure Design Guide
114	Performance-Based Design of Structural Steel for Fire Conditions
115	Pipe Ramming Projects

CONTENTS

PREFACE

This manual of practice was prepared by the Pipe Ramming Task Force of the ASCE Committee on Trenchless Installation of Pipelines (TIPS), under supervision of the Pipeline Division. The manual describes current pipe ramming practices used by engineers and construction professionals in designing and constructing pipelines under roads, railroads, streets, and other constructed and natural structures and obstacles. The TIPS Committee, under the leadership of Ahmad Habibian, P.E. (Past Chair) and Timothy Stinson, P.E. (Current Chair), is responsible for the efforts leading to this publication. The committee would like to thank contributors, task force members, and blue ribbon reviewers, whose names follow, for their support, time, and effort. The efforts of Baosong Ma, Professor, College of Engineering, China University of Geosciences, Wuhan, who was a visiting scholar at the Center for Underground Infrastructure Research and Education at The University of Texas at Arlington during the time this manual was being developed, is greatly appreciated.

Mohammad Najafi
ASCE Pipe Ramming Task Force Chair

ACKNOWLEDGMENTS

Contributors

Chapter 1—General Information
Kevin P. Nagle, TT Technologies
Collins Orton, TT Technologies

Chapter 2—The Planning Phase
Terry McArthur, HDR Engineering, Inc.

Chapter 3—The Design and Preconstruction Phase
George Davis, Missouri Department of Transportation

Chapter 4—Casing Materials
Jeff Bikshorn, Edgen-Murray Corporation
Dave Mittler, Permalok Corporation
Terry Moy, Woolpert Inc.
Mike Sechelski, Northwest Pipe Company

Chapter 5—Construction Phase
Fred Burlbaw, Hammer Head Mole
Hanson Turnbull, Hammer Head Mole

Chapter 6—Case Studies
Robert Carpenter, *Underground Construction* Magazine
Baosong Ma, China University of Geosciences
Collins Orton, TT Technologies

Task Force Officers
Mohammad Najafi, Chair
Larry Slavin, Vice Chair
Craig Camp, Secretary

Blue Ribbon Reviewers
Dennis Doherty, Jacobs
Ahmad Habibian, Black & Veatch Corporation
John Hair, J.D. Hair & Associates, Inc.
Mike Schwager, TT Technologies

ASCE Representatives
Matt Boyle, Manager, Book Production
Verna Jameson, Senior Coordinator, Technical Activities
John Segna, Director, Technical Activities

Pipeline Division Executive Committee
Ahmad Habibian, Ph.D., P.E., *Past Chair*
Michael T. Stift, P.E., *Chair*
Mohammad Najafi, Ph.D., P.E., *Vice-Chair*
Tom Iseley, Ph.D., P.E., *Senior Member*
Joe Castronovo, P.E., *Senior Advisor*
John Hair, P.E., *Secretary*
Terry McArthur, P.E., *New Member*
Tim Stinson, P.E., *Incoming Member*
Randy Robertson, P.E., *TAC Representative*

CHAPTER 1
GENERAL INFORMATION

1.1 INTRODUCTION AND BACKGROUND

This manual addresses pneumatic pipe ramming, which is widely used for installation of steel pipes and casings. The manual does not address other types of trenchless technology methods, which are also used for the installation of casings and pipelines, such as horizontal auger boring or horizontal directional drilling (see Section 1.5 for related documents).

Pipe ramming is a technique for inserting a steel pipe from a launch (drive) pit by means of ramming or pushing the pipe through soil using a pneumatic percussion hammer or rammer or a hydraulic jacking system. The pneumatic percussion hammer method uses the dynamic force and energy transmitted by the percussion hammer, which is attached to the end of the steel pipe, to drive the pipe through the soil. The hydraulic pipe jacking method uses the cyclical thrust of the hydraulic jacks to force the pipe through the soil. Steel pipe is used in essentially all applications because of its ability to withstand the ramming forces and typical loading conditions encountered. Steel pipe has been installed in lengths exceeding 400 feet (122 m) for diameters up to 60 in. (1,524 mm) and, for relatively short distances (e.g., 100 ft or 30 m), in diameters exceeding 120 in. (3,048 mm).

Pipe ramming is most appropriate for installation of large-diameter pipes over a short distance and for shallow-depth installations. It is suitable for all ground conditions except those for which the entire pipe perimeter is engaged with solid rock and is often successful where some other trenchless methods can lead to unacceptable surface settlement. Pneumatic pipe ramming is most commonly used in applications with a horizontal orientation, but it can also be used for vertical applications, such as driving piles or micropiles. The primary applications of this method are the placement of casings beneath roadways or railroads at crossings, or at crossings beneath streams. The casings may contain water or wastewater pipelines or serve as a utility gallery for electric, telecommunications, or fiber optic cables. The installed casing can also be used directly as a storm culvert or product pipeline in which interior coatings are applied after spoil is removed. Pipe ramming has also been used to install a larger pipe in place of an existing pipeline. Pipe ramming is an economical alternative to open trenching because it can reduce pavement damage, traffic disruptions, and the social costs associated with pipeline installations.

This manual of practice has been prepared to assist engineers, contractors, and owners who are using the pipe ramming method to design and execute pipe installation projects effectively, safely, and in conformance with project requirements and site conditions. The objective of this manual is to give a clear understanding of the method's capabilities and limitations; to outline important design and construction considerations; and to identify potential problems and prevention measures, thereby instilling confidence in the use of the method. The guidelines provided are based on case studies, workshops, project reviews, technical papers, and other information contributed by industry experts.

1.2 METHOD DESCRIPTION

The two major categories of pneumatic (or percussive) pipe ramming are closed-face and open-face.

The technique for closed-face pipe ramming (also called piercing, impact moling, or compaction method) includes welding a cone-shaped head to the lead end of the first segment of pipe (casing). The head penetrates and compresses the surrounding soil, and the casing is rammed forward. The soil–pipe installation interaction with this technique is similar to what occurs during soil-compaction methods. The soil is displaced and compressed around the outer surface of the steel pipe. This technique can be used for pipe up to 8 in. (200 mm) in diameter. This method is mainly used for installation of cables and small-diameter conduits under roads and streets. This method is widely used on nonengineered jobs.

Open-face pipe ramming allows the front end of the lead steel casing or conduit to remain open. With this technique, most of the soil particles remain in place, with only slight soil compaction occurring around the pipe during the ramming process. One advantage of open-face pipe ramming is the ability to swallow large rocks and boulders during the ramming process. Rocks and boulders as large as the inner diameter of the steel pipe can be swallowed by the casing. This technique is used for pipes larger than 8 in. (200 mm) in diameter.

To facilitate the open-face pipe ramming process, the lead edge of the first casing is usually reinforced by welding a steel band 12 to 14 in. (305 to 356 mm) wide around the exterior face of the pipe. The banding provides two advantages: (1) it reinforces the lead edge, and (2) it decreases the frictional drag around the casing. A steel band can also be welded around the inside edge of the lead section of pipe. This band adds further reinforcement, protects the casing end from damage if cobbles or boulders are encountered, guides the boulders into the casing for removal, and creates clearance for the soil within the casing. This clearance helps during the cleanout process and reduces the frictional drag within the casing. For small-diameter casings, after the casing installation is complete, the soil that has entered the casing can be removed by applying compressed air or water from either end. For larger diameters, augers can be used to mechanically remove the soil from inside the casing.

For either the closed-face or open-face process, bentonite or polymer lubricants may be needed for large-diameter pipes with longer ram lengths, or in some soil conditions, such as stiff clays or sands. A small steel supply pipe installed on top of the steel casing delivers the lubricant. The start of the supply pipe should be installed at a point approximately 24 in. (600 mm) behind the leading edge. The supply line is used to supply water, bentonite, or polymer lubricants outside or inside the casing (or both); to aid in spoil removal; to reduce exterior or interior skin friction; and to maintain the integrity of the hole being created.

1.3 RECENT INNOVATIONS (COMBINING PIPE RAMMING AND HORIZONTAL DIRECTIONAL DRILLING)

Some of the most interesting and impressive trenchless projects have been accomplished through the use of directional drilling. However, drill operators and manufacturers are finding new and creative ways of addressing particularly difficult projects and situations by using pipe ramming technology to assist directional drills. The techniques are changing the way drillers approach projects and respond to problems on the job.

Several techniques have been successfully used to help prevent failures in directional drilling projects or salvage bores by using pipe ramming technology. Thus, properly configured pneumatic pipe rammers can be used to salvage product pipes, remove stuck drill stems, and assist drills during product pullback, overcoming hydrolock, or other problems that may hamper the pipe installation. Drilling contractors throughout North

America have successfully used all of the directional drilling assist techniques described below.

1.3.1 Bore (Product Pipe) Salvage

A pneumatic pipe rammer may often be effectively used to salvage a stuck product pipe. For this procedure, a pneumatic pipe rammer is attached to the exposed end of the partially installed product pipe in an orientation that tends to pull the pipe from the ground. This can be accomplished through a fabricated sleeve (Fig. 1-1). A winch or other type of pulling device is used to assist the rammer during the pipe removal operation. The percussive power of the pipe rammer is often sufficient to free the stuck pipe and to then allow it to be readily removed from the ground.

1.3.2 Drill Stem Recovery

There are two possible configurations for applying a pneumatic pipe rammer to assist in drill stem recovery. Depending on the situation, contractors can directly pull the drill stem from the ground using the power of a ramming tool or, if the stem is still attached to the drill rig, they can use the ramming tool power to push on the opposite end of the stem to assist with the drill rig pullback action (Fig. 1-2).

1.3.3 Pullback Assist

The pullback assist technique helps install the product pipe in problematic situations. For example, when drilling underwater or in loose

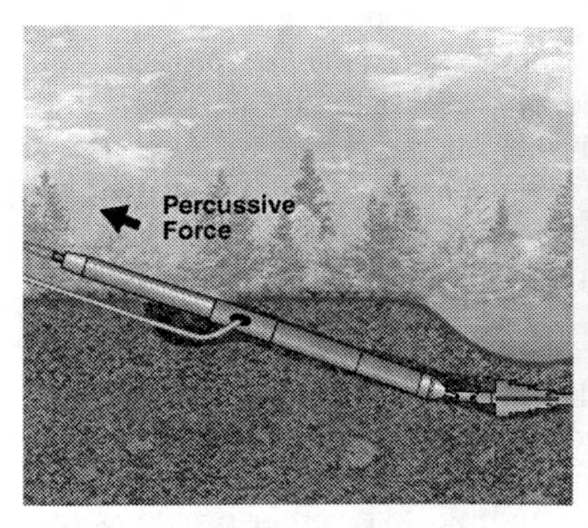

FIGURE 1-1. Bore salvage.

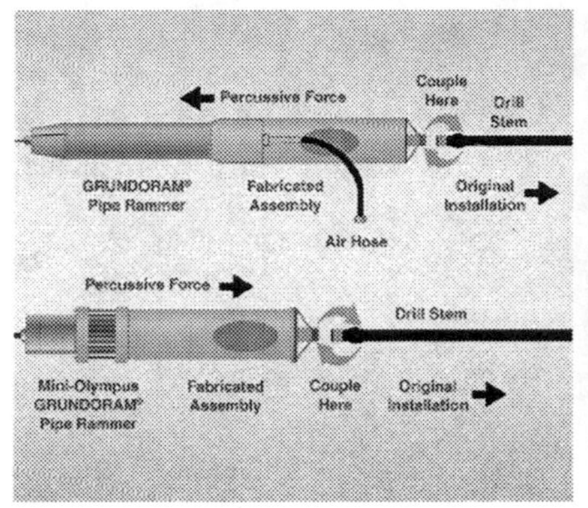

FIGURE 1-2. Drill stem recovery.

flowing soil conditions, or when there is loss of drilling fluid circulation, a condition known as hydrolock can occur. Hydrolock results when the pressure at the leading end of the product pipe restricts its forward movement. Alternatively, soil pressure along the side of the pipe due to partial collapse of the borehole may cause additional frictional drag, thus inhibiting the pipe movement. In such cases, the required pull forces may exceed the drill rig's pullback capability or the product pipe's tensile strength. The percussive pushing action of a pipe rammer applied at the tail end of the product pipe may be used to help free the immobilized pipe.

The pullback assist technique has been successfully used on steel pipe as well as high-density polyethylene. The technique can be used initially as a precaution in anticipation of possible problems, such as those described above, or after the pipe has become immobilized (Fig. 1-3). Response time, however, is a key factor. The rate of success greatly improves as the response time decreases. Therefore, many drilling contractors bring ramming equipment to directional drilling sites to be able to respond quickly to problems that may develop.

1.3.4 Conductor Barrel

The success of a drilling operation can often be determined before initiating the drilling operation. If the soil conditions at the planned entry point are problematic, the success of the entire project may be jeopardized. In such cases, the conductor barrel process may be appropriately used.

The conductor barrel technique differs slightly from the preceding methods because it is incorporated into the initial boring plan, rather than being deployed only in the event of a problem that may arise at some

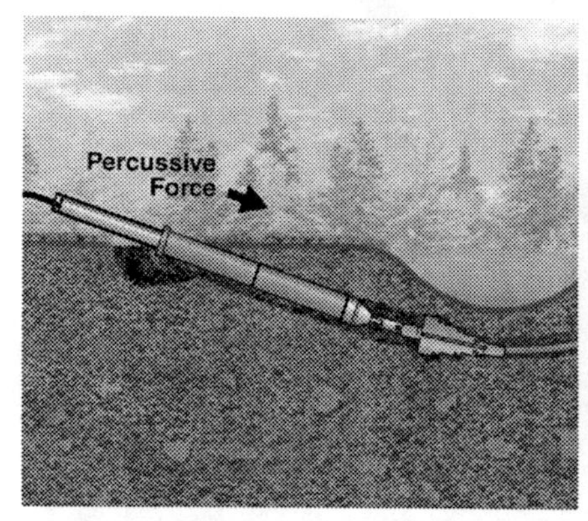

FIGURE 1-3. The pullback assist technique.

FIGURE 1-4. The conductor barrel process.

stage. In this method, a clear pathway is created through poor soil conditions, allowing the actual drilling operation to begin in more preferable soil conditions. Areas with loose, unsupported soils are prime candidates for the conductor barrel method.

During the conductor barrel process, open-face casings are rammed into the ground at a predetermined angle until desirable soil conditions are met. The spoil is removed from the casing with an auger or core barrel. Drilling then proceeds within the casing, beginning at a point where more desirable soil conditions are encountered. In addition to assisting drilling operations at the start of the installation, the conductor barrel can prevent situations in unstable soils in which drilling fluids under pressure force their way into waterways or wetlands, acting in a similar fashion to containment cells. The conductor barrel may also serve as a low friction section, facilitating pullback (Fig. 1-4).

1.4 ORGANIZATION OF THIS MANUAL

This manual of practice presents detailed information on pneumatic pipe ramming and hydraulic pipe jacking technology, supplemented by analysis of limitations and applications of these facts. Chapter 2 discusses the planning phase. Chapter 3 addresses the design and preconstruction phase, Chapter 4 provides information on the pipe materials, Chapter 5 discusses the construction phase, and Chapter 6 presents several case studies. At the end are a glossary and a list of references.

1.5 RELATED DOCUMENTS

American Society of Civil Engineers. (1996). *Pipeline Crossings,* ASCE Manuals and Reports on Engineering Practice No. 89. ASCE, New York.

American Society of Civil Engineers. (2002). *Standard Guideline for the Collection and Depiction of Existing Subsurface Utility Data,* CI/ASCE 38-02. ASCE, Reston, Va.

American Society of Civil Engineers. (2004). *Horizontal Auger Boring Projects,* ASCE Manuals and Reports on Engineering Practice No. 106. ASCE, Reston, Va.

American Society of Civil Engineers. (2004). *Pipeline Design for Installation by Horizontal Directional Drilling,* ASCE Manuals and Reports on Engineering Practice No. 108. ASCE, Reston, Va.

American Society of Civil Engineers. (2007). *Geotechnical Baseline Reports for Construction—Suggested Guidelines,* ASCE, Reston, Va.

American Society of Civil Engineers. (2007). *Pipe Bursting Projects,* ASCE Manuals and Reports on Engineering Practice No. 112. ASCE, Reston, Va.

Najafi, M. (2005). *Trenchless Technology: Pipeline and Utility Design, Construction and Renewal.* McGraw-Hill, New York.

Guidelines for Pipe Ramming. (2007). TTC Technical Report No. 2001.04, available at: http://www.latech.edu/tech/engr/ttc/publications/guidelines_pb_im_pr/ramming.pdf.

CHAPTER 2

THE PLANNING PHASE

The planning of a project has a major effect on the overall success because of the leveraging effect that decisions made at this phase have on all subsequent project activities. Planning is performed by all parties in different stages of the construction process, but initial planning is typically done by the project owner using the services of an engineer.

The nature of the project may have a great effect on the planning process itself. The planning phase for a planned and budgeted capital improvement can proceed in a more regulated and orderly fashion than that for an emergency pipe replacement caused by a collapse in a major roadway, i.e., an unexpected, unscheduled, unbudgeted, unwanted, and extremely urgent project. For an accelerated schedule appropriate for an emergency, the speed of completion often outweighs all other priorities. This may lead to the selection of a project delivery system and contracting method that focuses on speed, such as design-build, cost plus, etc., rather than preparation of bidding documents, including bid advertisement, opening, review, and award. These issues should be identified and considered in the first activity (background assessment) of the planning phase.

It is important in the planning process for the project owner to clearly identify, define, and communicate the project priorities to all those working on the project. The project owner should also perform a self-evaluation of his or her project management team to help ensure that these priorities are being followed and that conflicting priorities are not inadvertently introduced that may affect project performance.

Some suggested planning activities are listed in this section along with more detailed considerations for the various elements. The planning activities are listed in typical chronological order, which in general proceeds from the more general to the more specific concerns, with each activity relying, to some degree, on the previous activity's results. However, as discussed above, the project background and priorities may require modification, consolidation, rearrangement, or other changes to these suggested activities.

2.1 PLANNING ACTIVITIES

Proper planning helps to ensure that the project will meet the needs and priorities of the owner. The major activities that typically are performed during the planning phase include the following:

- background assessment,
- identification and screening of alternatives,
- data collection, and
- evaluation and selection of design alternatives.

A detailed discussion of each activity is presented below.

2.1.1 Background Assessment

The background assessment evaluates various factors, usually of a nontechnical nature, that establish the conditions and limitations of the project, including the project priorities. Some factors that may be considered include the following:

- Is the project a planned improvement or an emergency condition?
- What is the project budget limit?
- What is the project schedule?
- Are there any political considerations?
- What are the effects and ramifications to public safety and health?
- What are the constraints on construction (by any method)?
- What are the regulatory considerations and effects?

2.1.2 Identification and Screening of Alternatives

A general screening of potential alternatives based on the technical conditions and project needs serves to further eliminate candidate solutions and identify the type of preliminary design data that must be collected for the project to proceed. Some items that may be considered include the following:

- Installation of a new pipe, rehabilitation of the existing pipe, or repair of the existing pipe.

  - What are the existing and future land uses for the areas served by this pipe?
  - What are the current and future service requirements for this pipe?
  - What is the ability of this pipe to meet the identified service requirements?
  - Is the existing location of this pipe favorable to its continued use in the future?
- Construction method: conventional trenching vs. trenchless, or a combination of these methods.
  - What are the surface conditions?
  - What are the subsurface conditions?
  - Which alternative is more attractive overall: installation of a new pipe, rehabilitation of the existing pipe, or repair of the existing pipe?
  - What is the potential for settlement and other geotechnical issues?
  - Are there special safety considerations in the project area?
- Methods and materials based on present and future conditions.
  - What flow capacity is required?
  - What is the optimal diameter of the pipe for the identified service conditions?
  - What effects will the subsurface conditions have on the methods and materials being considered?
  - What are the pipe service conditions?
  - What are the line and slope tolerances?
- Any additional special considerations or data that may be required to successfully complete this particular project. For example, are there concerns about potential soil contamination?

2.1.3 Data Collection

This activity includes investigations to provide the information required to identify, screen, and evaluate candidate alternatives for consideration in the design phase. The purpose of data collection is to provide information relating to the following:

- Service conditions:
  - What are the existing and future service areas and land use?
  - What is the internal corrosion potential for the proposed pipe materials?
  - What are the structural requirements?
- Physical conditions:
  - What is the topography?

  - What are the surface features?
  - What and where are the existing utilities?
  - Should subsurface utility engineering be undertaken?
  - What and where are the sensitive areas within the project site?
  - What historical data are available for the project site?
- Subsurface conditions:
  - What are the general soil types, locations, and in situ conditions?
  - What is the potential for the presence of rocks, cobbles, or boulders?
  - Is groundwater present, and if so, what is its depth?
  - What is the soil's corrosion potential for the pipe exterior?
  - What is the soil's settlement potential?
- Possible soil contamination:
  - Is the soil contaminated?
  - Is special handling required to protect the equipment and work staff?
  - What are the requirements for disposal?

2.1.4 Evaluation and Selection of Design Alternatives

The last planning activity consists of the final screening and selection of the various alternatives to be addressed during the design phase. This final screening is based on the data collected above in addition to data regarding the following:

- Determination of the pipe service requirements and conditions:
  - What length of pipe is under consideration?
  - What are the structural requirements?
- Constructability and site limitation:
  - What are the surface effects of the construction techniques?
  - What are the easement needs of the construction techniques?
  - What effects will groundwater have?
  - What is the effect of other utilities?
  - Are there special restrictions on allowable noise levels?

The following sections examine in greater detail some of the factors listed above and their interrelationships.

2.2 PREDESIGN SURVEYS

Predesign surveys help to identify and obtain preliminary information needed for the design of the project. Although the responsibility for the surveys is not assigned to a specific party, it is preferable that the owner and the engineer obtain as much of this information as possible and incorporate it into the design, rather than require the contractor to collect such information. Information critical to the successful completion of the project can then be verified by the contractor in the field before beginning the work. The bidding and contract documents should delineate the extent to which the bidders (and ultimately the contractor) can rely on information supplied by the owner.

2.2.1 Land Use

The existing and future land use in the area affected by the line being installed or replaced is perhaps the most important detail that must be faced in planning capital expenditures. The land use will help identify the service conditions and service life of the pipeline (e.g., water or sewer). Such an obvious consideration has been overlooked in many projects. For example, installation of a culvert in an area that will undergo dramatic land use changes may soon have inadequate hydraulic capacity. This consideration may affect service conditions and design life expectations of the pipe under consideration. Appropriate state and local hydraulic design guidelines must be followed for pipe size selection.

2.2.2 As-Built Drawings

The as-built drawings and any available documentation from other projects in the vicinity of the project being considered should be reviewed. If previous soil borings are available, they should be reviewed and included with the supplemental information available to the bidders. Although extremely useful, the existing as-built drawings should not be assumed to determine the precise location of existing pipes, but they should be verified based on the other surveys described.

2.2.3 Site Conditions and Surface Survey

Site conditions, including surface improvements, must be thoroughly investigated and documented. The following details should be part of this documentation:

- the location of other existing utilities, both parallel and perpendicular to the line in question (see Section 2.2.4 of this chapter). This information must include the height or clearance limits of these other utilities.
- detailed notes showing surface improvements, such as trees and shrubs, type of ground cover, paving, and its condition. All pedestrian and motor vehicle traffic paths must be shown.

When any of the above details cannot be ascertained, the plans should be noted to reflect that the information is missing. All the information must be field verified by the contractor before beginning work.

2.2.4 Subsurface Survey

When designing and specifying trenchless construction, the existing underground site conditions (subsurface) should be investigated and documented to the degree appropriate for the technology selected. The following items may be considered for pipe ramming:

- subsurface soil conditions, including types of soils, rocks, and groundwater;
- existing buried utility locations, including types, sizes, depths, and crossing requirements (see Section 2.2.5 of this chapter);
- the potential for the presence of contaminated material; and
- other underground construction.

Some of the available subsurface technologies (such as ground penetration radar) may not work successfully in certain ground conditions. Proper information at the planning stage is essential for selecting the proper materials and methodology for the project.

2.2.5 Utility Locating

The predesign survey should locate and identify all of the existing underground utility lines in the area within 15 feet (5 m) laterally of the proposed pipe ramming operation. This effort can be accomplished by a variety of techniques, each yielding differing levels of information and certainty, including review of the as-built drawings available from the different utility owners, geographic information system (GIS) data, utility maps, pipeline markers, One-Call location and marking, carefully exposing the utility by excavation (usually by vacuum methods), and field surveying its location with respect to the project coordinate system. In some cases, additional field visits and exploratory efforts may be appropriate.

2.2.6 Subsurface Utility Engineering

The following discussion of subsurface utility engineering (SUE) is based on ASCE 2002.

Subsurface utility engineering can be defined as "a branch of engineering practice that involves managing certain risks associated with utility mapping at appropriate quality levels, utility coordination, utility relocation design and coordination, utility condition assessment, communication of utility data to concerned parties, utility relocation cost estimates, implementation of utility accommodation policies, and utility design." SUE offers an opportunity for a comprehensive and organized approach to the location and accommodation of existing underground utilities, which can provide more comprehensive information regarding existing utilities that may be present.

Current ASCE standards suggest the use of a "utility quality level which is defined as a professional opinion of the quality and reliability of utility information. Such reliability is determined by the means and methods of the professional. Each of the four existing utility data quality levels is established by different methods of data collection and interpretation." The four quality levels are the following:

- "Utility Quality Level A—Precise horizontal and vertical location of utilities obtained by the actual exposure (or verification of previously exposed and surveyed utilities) and subsequent measurement of subsurface utilities, usually at a specific point."
- "Utility Quality Level B—Information obtained through the application of appropriate surface geophysical methods to determine the existence and approximate horizontal position of subsurface utilities. Quality level B data should be reproducible by surface geophysics at any point of their depiction."
- "Utility Quality Level C—Information obtained by surveying and plotting visible above-ground utility features and using professional judgment in correlating this information to quality level D information."
- "Utility Quality Level D—Information derived from existing records or oral recollections."

ASCE (2002) provides an in-depth discussion of SUE. However, it is apparent that varying levels of existing utility location may be applicable depending on the specifics of the project. During the planning phase, the owner should decide what quality level of utility information is consistent with the project needs and risk management strategy, and the potential pitfalls of insufficient information. The SUE process offers additional potential benefits, including avoiding or reducing the following:

- conflicts with other existing utilities,
- delays in the construction schedule,
- additional construction costs, and
- inconveniences to the public.

The level of utility quality information is an important decision that should be made and executed as early as possible in the project.

2.3 ENVIRONMENTAL IMPACT AND BENEFITS

The owner and the design engineer must determine the acceptability of normal construction activity on the area that is being affected by construction operations. Construction activities can result in producing objectionable environmental side effects, such as dust, noise, and vibration. Dust levels can be controlled using water or chemical dust suppressants. Noise reduction may require additional noise control features on equipment, the use of smaller equipment, erection of sound barriers, appropriate work scheduling, or other measures.

The owner must have a good understanding and a realistic expectation concerning the materials and methods that have been selected to install or renew the pipe in the local community. The use of pipe ramming can yield significant benefits and reduce concerns with regard to the environmental impact. Pipe ramming requires only limited work locations with the ability to cross under a sensitive site without significant disturbance.

2.4 SOCIAL IMPACT AND BENEFITS

Trenchless replacement of underground infrastructure provides various social benefits:

- reduced pavement repair and cost;
- reduced likelihood of damage to surrounding utilities and structures;
- reduced disruption to vehicular traffic;
- enhanced safety by reduced amount, size, and duration of excavations;
- reduced magnitude of potential soil contamination and water pollution and associated control requirements; and
- more rapid restoration of service or operation of the facility.

2.5 PERMITS

In general, permits represent permission granted to the utility owner from the owning entity of an existing facility to construct or replace a pipeline under the existing facility. Permits typically specify construction conditions and conditions for continued occupation of the space beneath the existing facility (such as annual fees, maintenance requirements, reporting requirements, and insurance). Although the utility owner is held ultimately responsible by the entity issuing the permit for any violations, the contractor will also be responsible (to the utility owner) for compliance with permit requirements pertaining to the construction. Permits can require an extended processing time to obtain and therefore are typically obtained by the utility owner during the project design phase. The contract documents should list and contain a copy of the construction requirements for all permits obtained for the project. The contract should also require the contractor to adhere to the requirements of the permits.

Some typical locations and types of permits that may be required for a given buried utility project include the following:

- U.S. Army Corps of Engineers 404 discharge permit;
- wetlands crossing permits;
- floodplain development permits;
- stormwater control permits; and
- crossing permits for
  - city streets,
  - county roads,
  - state highways,
  - railroads, and
  - waterways.

Construction permits may be obtained from the following agencies or authorities:

- municipalities,
- state departments of transportation or railroad authorities,
- river authorities,
- counties,
- regulatory agencies,
- funding agencies, and
- private owners.

The required permits must be identified as early as possible during the project planning, and measures must be instituted to secure the permits in a timely fashion. Monitoring of existing structures (especially within 50 feet or 16 m of the pipe ramming location) may be needed if there is a concern on the part of the public agency or the contractor that the pipe ramming operation may negatively affect the existing structures.

2.6 JOBSITE LOGISTICS REQUIREMENTS

The owner must recognize and address community and neighborhood needs when specifying the following details of the contract documents, which will direct the contractor in his or her work:

- storage areas,
- equipment set-up areas,
- construction staging areas,
- location of major supporting equipment, and
- working hours.

This information should be incorporated into the contract requirements to avoid violation of private properties or unreasonable interference with any public or private operations. It is important to address the logistical space requirements associated with the anticipated construction techniques during the planning phase of the project. In some instances, the availability or lack of available space for logistics will be the controlling factor in selecting construction techniques. Additional details are provided in Section 5.1 of Chapter 5.

2.7 LENGTH AND ACCURACY OF INSTALLATION

The length of the pipeline segment that can be installed will be affected by the materials and methods that are selected. For pipe ramming, casing drive lengths may be limited by geotechnical conditions, equipment limitations, casing strength limitations, accuracy requirements, and cost factors. Typical ramming lengths are usually less than 250 feet (76 m). One of the limitations of pipe ramming is that its accuracy is almost entirely controlled by the initial setup. Once the ramming operating has begun, there is little that can be done to change the direction of the casing. For installations for which the slope of the completed installation is critical (i.e., for gravity sewers), the limitations of the ramming process may be compensated for by the installation of an oversized pipe casing that allows adjustment of an internal carrier pipe to overcome potential casing misalignment. It should be noted that some contractors have developed methods of influencing the final grade of the casing by installing external deflectors that act similar to aircraft flaps. This method requires a degree of trial and error due to soil variations and requires close monitoring of the casing during the installation. If soil conditions allow, a steering head can also be used, similar to the type used with auger boring machines. The steering head must be strengthened to withstand the impact loads associated with the pipe ramming operation. For additional information regarding steering of an auger boring head, see ASCE (2004).

2.8 GEOTECHNICAL CONSIDERATIONS

Potential settlement and surface heave will be greatly affected by the choice of the pipe material and the selected construction method. Pipe ramming technology has significantly advanced, allowing successful installation in relatively difficult ground conditions. For example, pipe ramming may be used in unstable soil conditions that may prevent the use of horizontal auger boring.

The project designer (or engineer) should fully understand the technology that has been selected, including its limitations and potential vulnerabilities. Existing ground conditions can change dramatically from one area to another and can significantly increase the difficulty of the installation. Closed-face pipe ramming develops a zone of displaced soils ahead of the casing. Open-face casings minimize soil displacement by allowing the soils to enter the casing pipe. However, if an unanticipated rock of sufficient size clogs the open end of the casing, it may lead to soil compaction, deflection from the design alignment, and potential heaving. The vibratory effects of the percussive ramming operation may cause soil consolidation and settlement in some instances. These factors emphasize the need for good subsurface information during the planning phase.

The potential applicability of various types of soils to pipe ramming is shown in Table 2-1. Table 2-2 supplements Table 2-1 as an aid in relating the standard penetration number (*N*) to pipe ramming capability. The standard penetration number is the number of blows required for driving a sampler through three 6-in. (152-mm) intervals. In general, pipe ramming may be readily applied for conditions with an *N* value of 20 or less. Conditions with *N* values greater than 20 will require pipe lubrication (see Section 5.4 in Chapter 5). An indication of the soil conditions may be determined by digging a hole (or holes) to the depth of the installation and

TABLE 2-1. Applicability of Pipe Ramming to Various Soils

Soil Type	Suitable for Ramming
Soft to very soft clays, silts, and organic deposits	Yes
Medium to very stiff clays and silts	Yes
Hard clays and highly weathered shales	Marginal
Very loose to loose sands (above the water table)	Yes
Medium to dense sands (below the water table)	No
Medium to dense sands (above the water table)	Yes
Gravels and cobbles (<4 in. or 100 mm in diameter)	Yes
Soils with significant cobbles, boulders, and obstructions (>4 in. or 100 mm in diameter)	Yes
Weathered rocks, marl, chalks, and firmly cemented soils	Marginal
Significantly weathered to unweathered rocks	Marginal

Source: Najafi 2005, with permission from McGraw-Hill.

TABLE 2-2. Identification of Soil Conditions Based on Standard Penetration Number (*N*)

N Value	60	40	20	10	6 or Less
Strength/ firmness	Very hard or very weak	Very stiff	Stiff	Firm	Soft
Penetration	Easily broken by hammer. Penetration to about 5 mm with knife	Easily broken by thumbnail. Penetration to about 15 mm with knife	Indented by thumb	Penetrated by thumb with effort	Easily penetrated by thumb

manually pushing against the soil with a person's thumb and determining the relative difficulty in penetrating the soil, as indicated in Table 2-2. In rocky conditions, including hard cobble, rocks smaller in size than the internal diameter of the casing may be engulfed by the pipe, pushed aside, or possibly cracked. Pipe ramming has performed well in conditions including gravel, soft limestone, glacial till, shale, and caliche.

The determination of the soil conditions will be of assistance in selecting the proper casing (see Section 3.2 in Chapter 3) and equipment (see Section 5.3 in Chapter 5).

2.9 WORKING (DRIVE AND RECEIVING) PITS

Working pit locations (both drive and receiving pits) are affected by the casing pipe size and methodology selected to perform the work. The pipe ramming function is performed at the entrance (or drive) pit. The casing pipe exits the ground at the receiving pit after being driven under the crossing. In constructing and maintaining the drive and receiving pits, the project designer should consider the following:

- the location of the various utility companies' facilities,
- street rights-of-way and other utility easements (crossing permits typically limit the effect on traffic operations),
- subsequent repair of damage to the property of other parties as a result of the pipe ramming process, and
- trench shoring or bracing in compliance with OSHA standards.

Additional details and requirements for the pits are provided in Section 5.1.2 of Chapter 5. In relatively long or difficult installations, additional pits may be required.

2.10 COST CONSIDERATIONS

Engineers and utility owners must evaluate the specific characteristics of each project and select the method or methods that can most safely and cost-effectively fulfill project objectives and requirements. The selection of the construction methods or options should not be determined exclusively by contractors because contractors are often optimistic about their capabilities and may desire to execute a project for a variety of reasons, including a present lack of contract work and corresponding availability of specific equipment and crews. However, an understanding of the capabilities and limitations of the equipment and materials, and the constraints of ground conditions in combination with contractor experience, are key factors in implementing a successful pipe ramming installation. Because pipe ramming technologies and procedures continue to evolve with improved equipment and materials as well as field experiences, material and equipment manufacturers and experienced contractors should be consulted to optimize the application for each project. In general, in all instances consulting and partnering with experienced and reputable contractors can assist in determining the optimum solutions and overall minimal project cost.

Construction project costs are based on direct costs (labor, materials, and equipment), indirect costs (field overhead and head office overhead), and markup (contingency and profit). Because trenchless technology methods are relatively new, the historical costs may not be as available as those for conventional open-cut methods. Furthermore, many midsize or small communities may have difficulty identifying quality and experienced contractors submitting bids for trenchless technology projects, and specifically for pipe ramming projects. It is recommended that the project planner contact other facility owners who have used pipe ramming for similar projects to discuss their cost experiences.

Although not included in the above cost categories, the "social costs" of utility construction represent an important consideration. Social costs include those directly or indirectly associated with public inconvenience; traffic disruption; short- or long-term damage to pavement, existing utilities, and nearby structures; noise and dust; loss of business; potential safety hazards; and damage to the environment and green areas. Trenchless technology methods, including pipe ramming, reduce or eliminate most of these social costs, often rendering these techniques more cost-effective than conventional open-cut trenching methods. Contractors are typically not concerned about social costs and do not reflect these costs in their bids. Social costs should therefore be considered by owners and engineers as part of a "lifecycle cost" analysis during the project planning phase.

In a conventional open-cut project, the majority of construction effort (approximately 70%) is expended for reinstatement of the ground surface. With trenchless methods, such as pipe ramming, a "better quality" pipe installation may be accomplished because most of the construction attention and efforts are directed toward the new pipe material and its installation. This tends to eliminate or minimize future operation and maintenance expenses, reducing the overall lifecycle cost of the project.

CHAPTER 3

THE DESIGN AND PRECONSTRUCTION PHASE

3.1 FEASIBILITY AND RISK ASSESSMENT

The pipe ramming method can be used in a wide variety of soil conditions ranging from sand to clay. Cobbles and boulders do not represent significant problems. However, conditions such as groundwater, flooding, and potential obstructions (e.g. solid rock, existing utilities, man-made obstructions, and unexpected geological conditions) may pose a challenge to pipe ramming projects. It is beneficial to analyze the suitability of the pipe ramming method under anticipated site conditions.

3.2 STEEL CASING DESIGN

Steel pipe is essentially the only type of casing used in the pipe ramming process. The pipe should be new, of good quality, and well prepared. Machine-cut beveled ends ensure casing straightness (alignment), joint-end squareness, and effective transfer of force from the pipe ram to the steel casing. Short lengths of pipe may be driven when the working area (both the drive and receiving pits) is limited, but longer lengths of pipe are preferred to minimize overall installation time and maintain a higher line and grade accuracy. If closed-face pipe ramming is used, a symmetrical cone or wedge is welded onto the lead casing section. For open-face pipe ramming, a steel reinforcing band is welded at the front of the lead section of casing. Closed-faced ramming may be used for casing diameters up to 8 in. (200 mm) outside diameter. Larger diameter casing requires open-face ramming techniques.

3.2.1 Banding the Casing

As described in Chapter 1, the use of a leading-edge outer band on an open-face casing is recommended when ramming in all soil conditions, especially on a relatively long segment. This band compacts the soil, relieves pressure on the casing by decreasing skin friction, and acts in conjunction with the leading edge of the pipe to break up boulders and cobbles around the pipe and allows the broken fragments to pass through. It also may act as a guard for a bentonite slurry fitting, if required, at the leading edge of the pipe.

The design and placement of the band is generally based on contractor experience, but it is usually 12 to 14 in. (305 to 356 mm) wide around the exterior face of the pipe and is rolled to fit the outer diameter of the casing. The band is placed so that it leads the casing by approximately ½ in. (12 mm) and is welded securely along the front on its inside surface and along the rear on its outside surface. Banding the casing is the first operation in an open-face pipe ramming installation.

3.2.2 Casing Sizes

The typical sizes of casings range from 4 to 96 in. (100 to 2,440 mm) or greater, depending on the project requirements.

3.2.3 Casing Wall Thickness

The thickness of the casing depends on the overall pipe diameter. Table 3-1 provides recommended wall thicknesses for commonly encountered pipes up to 80 in. (2,000 mm) in diameter. Thicker sidewall or increased strength steel may be required if excessive jacking forces are anticipated due to ground conditions or when crossing under highways, railroads, or when required by permit. Thicker pipe may be desired to account for possibly greater than anticipated push forces or if pipe is to be installed at a large or a shallow depth directly under a highway or railroad. The pipe wall thickness does not include the leading edge band on the pipe. The pipe ramming equipment manufacturer should be consulted for pipes greater than 80 in. (2,000 mm) to recommend an optimum pipe wall thickness.

Information concerning the design of the pipe, including casing wall thickness, to account for possible installation and service loads may be found in Watkins and Anderson (2000) and also Moser (2001).

3.2.4 Midweld

A midweld is sometimes used for assembling casings of 48 in. (1,200 mm) and larger diameters. Midweld is two or more pipe joints

TABLE 3-1. Recommended Pipe Thicknesses for Commonly Encountered Steel Casing Sizes Used in Pipe Ramming

Outside Diameter of Pipe		Minimum Wall Thickness			
		Bores up to 65 ft (20 m)		Bores exceeding 65 ft (20 m)	
in.	mm	in.	mm	in.	mm
6	150	0.25	6.3	0.27	7.1
8	200	0.25	6.3	0.27	7.1
10	250	0.25	6.3	0.27	7.1
12	300	0.25	6.3	0.27	7.1
14	350	0.27	7.1	0.31	8
16	400	0.27	7.1	0.31	8
18	450	0.31	8	0.39	10
20	500	0.31	8	0.39	10
24	600	0.39	10	0.47	12
28	700	0.39	10	0.47	12
30	750	0.47	12	0.55	14
32	800	0.47	12	0.55	14
36	900	0.47	12	0.62	16
40	1,000	0.47	12	0.62	16
42	1,050	0.59	15	0.62	16
48	1,200	0.59	15	0.7	18
51	1,300	0.62	16	0.7	18
55	1,400	0.7	18	0.78	20
60	1,500	0.75	19	0.87	22
72	1,800	0.87	22	1	25
80	2,000	0.875	22.2	1	25

Source: TT Technologies, with permission.

welded to form a longer pipe section. The weld should be properly made to avoid leakage and infiltration.

3.2.5 Installation Loads

The casing should have sufficient strength to withstand the axial and transverse loads due to pipe–soil interaction during and after the installation, as well as the repeated hydraulic or pneumatic push forces applied by the pipe ramming equipment. Lubrication applied to the exterior of the casing may be effectively used to reduce the required push forces and overall installation loads (see Section 5.4 of Chapter 5).

3.2.6 External Loads

Because the pipes installed by the pipe ramming method often pass under highways and railroads, there may be a considerable amount of external service loading on the pipe. Such factors should be considered when selecting the wall thickness and the strength of the casing, especially if the pipe is to be placed at a shallow depth directly under such traffic conditions. It is assumed that the installation will comply with all local, state department of transportation, and railroad guidelines and requirements regarding depth of cover.

3.2.7 Steel Casing Corrosion Protection

In general, the casing thickness that is used in pipe ramming is sufficiently thick that corrosion protection coatings are not required. Because coating on the casing pipe may be damaged during installation from the repeated action of the pipe ramming equipment or from the casing being pushed through granular soil, coatings are not typically used. In extremely corrosive environments, it is therefore recommended that an increased wall thickness be used to provide greater longevity. Recently, however, epoxy coatings have been used successfully for gas transmission pipes and may be implemented for difficult pipe ramming operations. The epoxy coating is generally able to withstand the abrasive effects of the installation and decreases the friction between the steel pipe and the soil, thus facilitating the operation. The epoxy coating could also provide additional protection in corrosive soil environments.

3.3 CARRIER PIPE DESIGN AND INSTALLATION

If necessary, after successful installation of the casing pipe, an internal carrier pipe can be installed. Carrier pipe is installed by attaching wooden skids or premanufactured casing spacers to the carrier pipe before assembly. The carrier pipe is then installed, one piece at a time, from either the entry or exit pit. It can be installed by pushing by hand, with a boring or jacking machine, by pulling with a winch, or by other possible methods. Unlike the steel casing, the carrier pipe can be of various materials (e.g., steel, HDPE, PVC, concrete, or fiberglass).

3.3.1 Blocking Spacers

For gravity sewer installations, where installation of the carrier pipe on line and precise grade is necessary, it is extremely important that the carrier pipe be restrained to prevent flotation. This can be accomplished by using a differential wood blocking banded to the carrier pipe to allow for adjustment of the grade inside the casing. Alternatively, pre-manufactured spacers or casing insulators may be used. These spacers are manufactured in plastic, fiberglass, stainless steel, and carbon steel. They can also be coated in epoxy, rubber, and various other materials. Manufacturers provide recommendations for the design and spacing of the spacers. If the carrier pipe is properly supported with spacers, it may not be necessary to fill the annular space between the carrier pipe and the casing (see Section 3.3.2 in this chapter). This arrangement would also facilitate the subsequent removal of the carrier pipe if future maintenance becomes necessary.

3.3.2 Internal Casing Grouting or Filling

After the internal carrier pipe is blocked or supported inside the casing, grout or sand backfill can be used to fill the annulus of the pipe. Lightweight, low-strength cellular grout and possibly bentonite grout are excellent products for this application. Cellular grouts have a lower density than common sand–cement grouts, reducing the tendency for flotation of the carrier pipe. They also have superior fluidity, allowing for low installation pressure. For most applications, a compressive strength of 150 $lb/in.^2$ (10,342 kPa) is adequate. A disadvantage of the sand–cement grout is the heat of hydration, which may cause damage to small plastic pipes in a large annulus.

Because sand and pea gravel alone do not have sufficient capability to restrain the carrier pipe from floating or moving, the use of blocking spacers to hold the pipe down is important. This is especially true with larger diameter pipes. Carrier pipes may displace sand and pea gravel even if the annular space between the carrier and the casing is entirely full. In addition, sand and pea gravel are abrasive and may damage the pipe. The installation process uses an air jet to place the material inside the casing under high pressure. It is not uncommon to damage joints or to blow holes in pipes. Smaller casings that do not allow personnel entry are very difficult, if not impossible, to fill completely.

The alternative of using properly designed spacers eliminates the need for grout or backfill, thereby reducing the risk of damage to the pipe. Furthermore, the use of any type of grouting in the annular space essentially eliminates the possibility of removing the carrier pipe for future maintenance, if necessary. However, if spacers are used, there is still the possibility of accessing the pipe for maintenance or removal.

3.3.3 External Casing Grouting

In sandy or unstable soil conditions, there is a possibility of void formation in the line of the bore. In this condition, grouting outside the casing is strongly recommended, using approved grouting materials and methods.

3.3.4 Hydrostatic Pressure

The internal and external hydrostatic pressure experienced by the carrier pipe must be considered in its design. An appropriate wall thickness must be selected, consistent with the inherent material strength.

3.3.5 Corrosion Protection

Sufficient corrosion protection treatment should be given to prevent the carrier pipe from degrading. Several types of pipes are available that are highly resistant to corrosion.

3.4 CONTRACT DOCUMENTS

Pipe ramming projects are frequently designed to accomplish a critical crossing or interconnection and are typically included as part of a larger pipeline construction project. As a result, the contract documents generally address the pipe ramming portion of the installation as part of the overall project. Although the general categories of items to be presented in the specifications and drawings for the pipe ramming portion are similar in many ways to pipeline construction in general, the success of a pipe ramming project requires that its particular construction considerations be specifically addressed.

It is important that the contract documents be complete, clear, and concise and that they be prepared by individuals with experience in the specified methods and technology. These documents may include, but are not limited to the following:

- the scope of work and special conditions,
- drawings,
- technical specifications, and
- geotechnical information.

3.4.1 The Scope of Work and Special Conditions

The contract documents should provide information about the scope of work and special conditions of the job. The general conditions include the type of job, amount of work, and other relevant data for the contractor. Special conditions include information about specific tasks or characteristics about the job (see Section 2.1 in Chapter 2). Any other information available about the jobsite should be included in the contract documents.

3.4.2 Drawings

Drawings for the work to be performed should provide clear information about existing jobsite conditions, including constraints, and the construction to be performed, including depth of installation and all critical clearance requirements. The drawings represent the most effective means to communicate special conditions and details related to the project. In particular, the drawings for the pipe ramming project should provide the following:

- limits of work (horizontal and vertical control references);
- topography and plane metrics and survey points of existing structures;
- boundaries, easements, and rights-of-way;
- existing utilities, sizes, locations, and construction materials;
- the plan and profile of the design installation and alignment;
- the location and size of the launching and receiving pits;
- material and equipment layout and storage areas;
- details for connections to the existing piping system; and
- restoration plans.

Drawings may also include information regarding erosion and sediment control requirements, storm water flow bypassing plans, service connection and reinstatement details, and support activities to acquire permit approval, such as jobsite layout and pit location.

3.4.3 Technical Specifications

Technical specifications supplement the drawings in communicating project requirements to the contractor and include additional detailed construction or material requirements. These specifications may include casing and carrier pipe materials and construction considerations.

3.4.3.1 Casing and Carrier Pipe Materials. The following requirements may be specified by the contractor:

- standards and tolerances for casing materials, diameter, wall thickness or class, or testing and certification requirements;
- the pipe manufacturer's experience or history with similar work;

- construction installation instructions for casing handling, installation, and pipe joining;
- fittings, appurtenances, connections, and adaptors;
- requirements for annular space grouting; and
- carrier pipe requirements (e.g., materials, diameter, or wall thickness).

3.4.3.2 Construction Considerations. The following may be included as submittal requirements by the contractor:

- traffic control requirements;
- erosion and sediment control requirements;
- documentation for existing conditions (e.g., photographs, surveys, video, or interviews);
- pit details and earth support systems;
- if necessary, plans for ground movement monitoring and existing structure protection;
- accuracy requirements of the installed casing or carrier pipe (line and grade);
- daily construction monitoring reports;
- blocking or spacers for carrier pipe installations;
- field testing and inspections;
- requirements for pipe joining, pipe leakage, disinfection, backfill, annular space filling (e.g., grouting); or
- site restoration and spoil material disposal requirements.

3.4.4 Geotechnical Information

All geological and geotechnical information available to the owner should be detailed in the geotechnical report. Information about the type of soil, groundwater conditions, and water table are to be detailed in the geotechnical information.

3.4.5 Differing Site Conditions

Preliminary investigations, such as utility mapping and geotechnical surveys, should minimize many of the common obstructions occurring during the pipe ramming operation and should be identified in the geotechnical information provided to the contractor. Nonetheless, it is possible that unexpected conditions may be encountered during the actual construction phase. Changes in soil conditions, such as extremely hard and large rock boulders or changes in groundwater, can create situations that can halt a pipe ramming operation. Water in a boring operation can have major consequences. A preliminary survey, immediately before construction begins, should be conducted on the site to investigate and confirm the existing water table conditions before commencing work on the project. Proper dewatering techniques must be implemented to pump out water. Note that water table conditions may change at different times of the year.

A risk-sharing strategy, such as one recommended by ASCE (2007), is suggested. Usually, soil conditions (i.e., rock, clay, sand, or water table) when reasonably described, indicated, or implied in the contract documents, are not grounds for a change of conditions claim when using pipe ramming. However, if site conditions are significantly different than those described in the contract documents and the contractor can show that the different conditions affected the work, the contract value should be adjusted accordingly. Valid claims for differing underground conditions should generally be limited to unknown items (e.g., natural and man-made obstructions). The contract documents should contain a clause requiring written notification of claim for change of conditions within a certain number of days from the date of first discovery, after which no claim would be allowed.

Some situations that can create differing site conditions are the following:

- boulders larger than the pipe to be installed or extremely difficult to ram through,
- mixed ground conditions (soft ground changing to rock), or
- contaminated soils.

3.4.6 Existing Utilities

Existing underground utilities should be located and exposed before proceeding with the work. The contract should clearly identify responsibilities of the owner and the contractor in locating existing utility lines. Damage to adjacent utility lines is a major concern to any underground boring operation and should be avoided at all costs. The financial liability for damages to existing utilities or other structures or facilities may be of major consequence.

3.4.7 Dispute Resolution

Because there is a possibility of encountering unforeseen conditions during an underground construction project, it is imperative that a well-devised dispute resolution plan be included in the contract document. For example, a "differing site conditions" clause, consistent with the above discussion, should be included in the contract to allow a mechanism for resolving issues related to differing site conditions in a timely manner. It is in everyone's best interest to resolve any conflicts quickly, fairly, and equitably.

3.5 CONTRACTOR PREQUALIFICATION AND SUBMITTALS

Typically, the general contractor subcontracts the pipe ramming operation, and the owner or government agency does not directly retain the ramming subcontractor. It is recommended that the general contractor check the background, experience, and reputation of the subcontractor and the operator to ensure that he or she meets the requirements of the contract documents. The contract can require that the contractor shall seek the concurrence of the owner regarding the subcontractor's qualifications meeting the requirements of the contract documents before awarding the subcontract.

3.5.1 General Information and History

Construction companies are generally required to provide information on the type of organization, types of work undertaken, and type of personnel involved in the project during the first part of the program, including the following:

- the name and type of organization (e.g., corporation or partnership);
- a list of affiliate, partner, or subsidiary companies;
- the organizational setup within the company;
- the type of contracting business;
- the type of projects usually undertaken;
- the type of work performed using their own crew; and
- the history of change order requests.

3.5.2 Experience and Equipment

The experience of the personnel and the experience of the firm as a whole are considered as part of the prequalification requirement. Contractors are commonly required to provide information on the type and quantity of the equipment owned by the firm and the capacity and durability of that equipment. The following information may be requested:

- a list of projects completed;
- the completion rate of awarded projects;

- experience, history, certification, training, and references of the contractor;
- key construction personnel and their qualifications;
- a list of equipment owned (including name, description, quantity, and capacity);
- relevant materials available; and
- a field safety plan and safety records.

CHAPTER 4

CASING MATERIALS

Similar to other pipeline applications, material considerations for the casing reflect a combination of criteria based on the particular site conditions, installation requirements, and the utility application; i.e., the product material or service to be conveyed by the carrier pipe. Table 4-1 provides reference material standards for general applications (see also Section 4.2.3 of this chapter). Pipes with interlocking press-fit end connections are available in seamless, straight seam electrical resistance welded, double-submerged arc welding rolled and welded with applicable size range for ramming from 6 5/8-in. (168-mm) outside diameter to 144-in. (3,658-mm) outside diameter in various thicknesses. If pipes with interlocking press-fit end connections are used as carrier pipes, joint connections should include a double O-ring gasket, and fluid pressure should be limited to a maximum of 300 lb./in.^{2} (2,068 kPa).

4.1 PHYSICAL PROPERTIES

4.1.1 Casing Diameter

For typical applications in which the pipe serves as an outer casing for an internal carrier pipe or group of ducts, the size of the carrier pipes, spacers, and clearance requirements dictates the minimum internal diameter requirements. For those applications for which the casing itself serves as a product pipe, such as a storm water culvert, the hydraulic flow capacity may represent the primary criterion for minimum size selection.

In some cases, a larger casing than the minimum diameter otherwise required for the utility service may be required. If relatively large cobbles or small boulders are present, the casing diameter may need to be increased

TABLE 4-1. Reference Standards

Standard	Usage
ASTM A53, Grade B or better	a. available as seamless or straight seam electric resistance welded b. applicable size range for ramming from 6 5/8-in. (168-mm) outside diameter to 26-in. (660-mm) outside diameter in various wall thicknesses c. suitable for all applications
ASTM A106, Grade B or better	a. available as seamless only b. applicable size range for ramming from 6 5/8-in. (168-mm) outside diameter to 48-in. (1,219-mm) outside diameter in various wall thicknesses c. suitable for all applications
ASTM A139, Grade B or better	a. available as straight seam electric fusion welded, rolled and welded, or spiral welded pipe b. applicable size range for ramming from 6 5/8-in. (168-mm) outside diameter to 144-in. (3,658-mm) outside diameter in various thicknesses up to 1 in. (25 mm)
ASTM A252, Grade 2 or better	a. available as straight seam electric resistance welded, electric fusion welded, seamless, spiral welded, or rolled and welded b. applicable size range for ramming from 6 5/8-in. (168-mm) outside diameter to 144-in. (3,658-mm) outside diameter in various wall thicknesses c. suitable for all applications except as carrier pipe (spiral welded pipe is not recommended for ramming)
API-5L, Grade B or better	Alternate standards
API-2B, Grade B or better	Alternate standards
AWWA C200, 36 ksi (248 MPa) minimum yield	Alternate standards

to avoid potential obstructions and allow their removal with other spoil material through the casing. Additional clearance for access may also be warranted to facilitate spoils removal and help ensure a successful installation.

4.1.2 Casing Wall Thickness

The wall thickness must be sufficient to allow the casing to withstand the combination of loads to which it will be subject during the installation and anticipated design service life. Based on the required diameter, the dead load from the soil above the pipe and any live loads from the highway, railroad, or aircraft must be considered. The type of soil (e.g., sand, clay, gravel, or with the presence of cobble or boulders) and groundwater conditions are also important. The maximum hammer energy and forces that will be applied to the pipe, based on the length of the drive section and possible use of lubricants to reduce required push forces, will be important factors. In many cases, governing bodies, such as the highway department, railroad, or municipality, specify minimum wall thickness requirements for a given diameter, which must be considered in the selection process.

4.1.3 Casing Individual Length

The maximum allowable length of the casings installed is most often governed by the size of the workspace. If the pipe is being pushed from a pit or shaft, it will necessarily have to be installed in individual lengths that can be accommodated by the pit or shaft layout, in contrast to an at-grade installation, where available space may not be a constraint.

The maximum casing length may affect the type of pipe joining method to be used. If field butt-welding is required, shorter lengths of pipe will require more welded joints and a correspondingly greater downtime for welding. If interlocking press-fit connections are used (see Section 4.2.3 of this chapter), shorter lengths are more feasible because the downtime would be greatly reduced. The maximum length of a steel pipe section is 50 ft (16 m), with common available maximum lengths of 40 ft (13 m) and typical lengths of 10 ft and 20 ft (3.3 and 6.5 m).

4.2 AUXILIARY PROPERTIES OR FEATURES

4.2.1 Coatings and Linings

The use of exterior coatings on the casing is generally not recommended because of the likelihood of damage at the pipe–soil interface. However, after the installation of the casings, a variety of interior coatings and linings (e.g., liquid epoxy, polyurethane, or cement mortar) may be used for protection of the internal carrier pipe. Because of the potential for abrasion and damage from contact with spoils during the ramming process, proper surface preparation and application of coatings should be performed after the spoils are fully removed.

4.2.2 Fittings, End Treatment, and Appurtenances

Steel casing pipe can readily be provided with various fabricated or supplied fittings that maximize field productivity. As described in Section 1.2 of Chapter 1, the leading edge of the first casing is usually reinforced with a steel band (or shoe) welded into place on the outside, and possibly inside, surface of the pipe (Fig. 4-1). The leading edge of the outer cutting shoe should be set back behind the leading edge of the casing, and the inside band should be recessed back from the leading edge of the casing as well. When the cutting shoes have been properly placed, the soils will be funneled into the casing, where they can be removed as required.

If the installation is to be lubricated during the ramming operation, threaded ports can be provided in the pipe. These threaded ports can also function as ports for postinstallation grouting of the exterior of the pipe, if required. In the case of press-fit connections, a reinforced push ring machined to fit inside the female end of the pipe can be provided by the manufacturer.

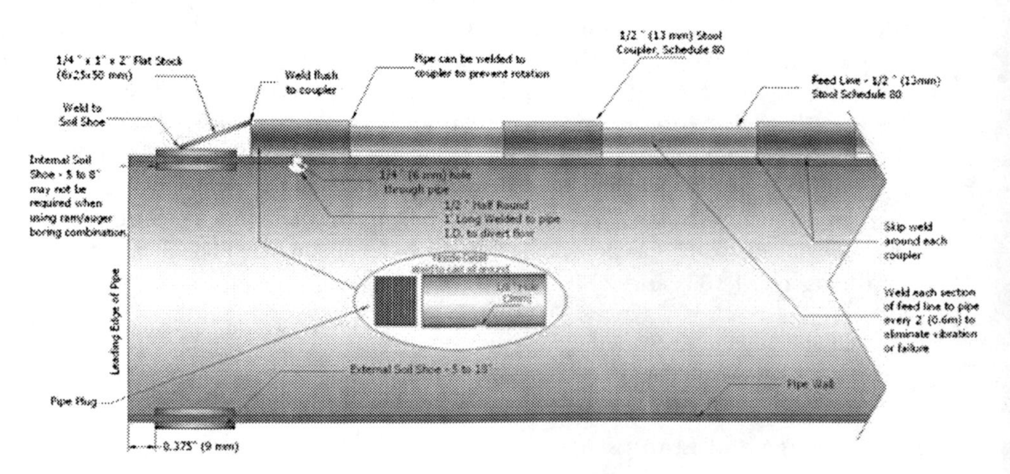

FIGURE 4-1. Typical lubricating line installation.

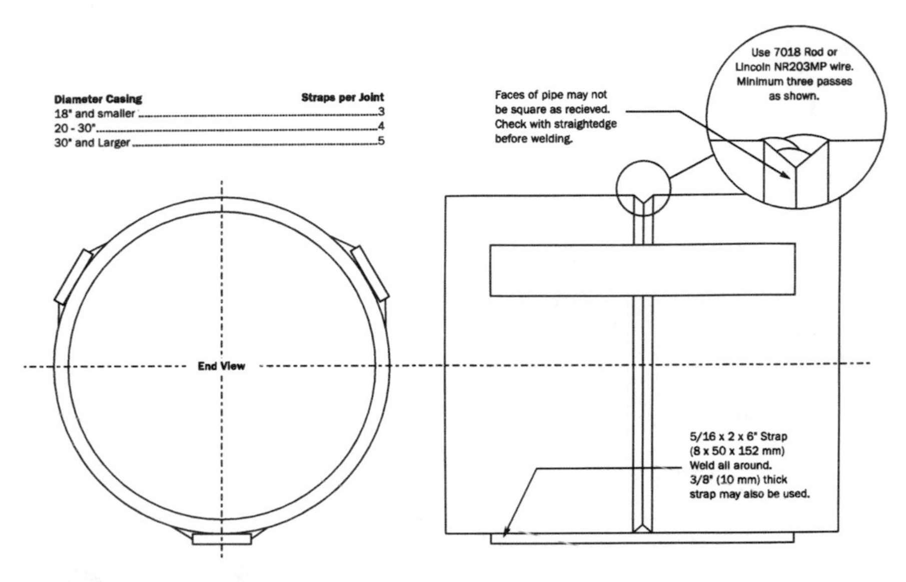

FIGURE 4-2. Joining multiple casings.

The casings can also be provided with lifting lugs for ease of handling, chain eyes for affixing the ramming cone (for closed-face applications), or other features or appurtenances to facilitate handling and field joining. Fig. 4-2 illustrates details of ends provided on each casing section to facilitate field welding and joining of multiple casings.

4.2.3 Interlocking Press-Fit Connections

Permalok Corporation supplies a special press-fit interlocking product that is available in seamless, straight seam electric resistance welded (ERW), double-submerged arc welding (DSAW) rolled and welded versions. It is available in sizes ranging from 6 5/8-in. outside diameter to 144-in. outside diameter (168 to 3,658 mm), and in wall thicknesses of 0.250 in. to 2 in. (6 to 51 mm). Although most often used in a casing pipe application, with the addition of a double O-ring gasket, it can also be used as a carrier pipe capable of handling up to 300 lb/in.^{2} (2,068 kPa) of internal pressure. Product standards and specifications can be obtained from the manufacturer (www.permalok.com).

CHAPTER 5
CONSTRUCTION PHASE

5.1 WORKSPACE

The workspace must provide sufficient room for the safe operation of the equipment and for storing materials. Workspace includes both the space available at the site and the space inside the pit. The space inside the pit should allow for free movement of personnel and equipment. The pit should be large enough to permit the personnel to work around the casings during the ramming operation and to attach and detach pipes. The space should also be large enough for the crane or excavator to lower the new pipes into the pit and to remove spoil from within the casings. The jobsite layout should be well planned (see Section 2.6 of Chapter 2) to facilitate storage of various materials and equipment required for the job, as well as to address the location and size of the working pits (see Section 2.9 of Chapter 2).

As mentioned in Section 2.5 of Chapter 2, local authorities should have been notified during the planning phase and all required permits should already be obtained. If necessary to venture into adjacent properties for pit excavation or workspace around the work pit, appropriate permissions to do so should be obtained and possibly justified with proper compensation.

5.1.1 Jobsite Layout

The jobsite layout should be prepared well in advance, before commencement of work on the site. The layout should provide adequate space for storage of equipment, material, and job trailers. The material required for the job, including pipe rammers, collets (holding devices), augers, casings, carrier pipes to be installed, and lubrication systems, should be close to the work pit. Material storage should be well planned to facilitate handling and to reduce travel time. The equipment required for the job, including the generator to supply power and the welding machine, should be close to the launch pit.

Other equipment, such as a crane or an excavator, should be placed with sufficient room to maneuver around the launch pit so that the various pieces of equipment do not interfere with each other during operation. Spoil-removal facilities and techniques should also be taken into consideration during the planning phase. If using compressed air to remove spoil from inside the pipe (typically used for casings up to 30 in., or 762 mm, outside diameter), care should be taken to avoid blowing the spoil toward populated areas.

A typical jobsite layout should be compatible with present easements and rights-of-way, and should consider the following:

- storage of materials (e.g., pipe sections, pipe rammer, and bulk materials for making lubrication fluids);
- lubrication systems; and
- support equipment (e.g., backhoes, forklifts, cranes, and air compressors).

5.1.2 Pits

A pipe ramming operation requires two work pits, the drive (launch) pit for performing the ramming operation and the exit pit for receiving the lead end of the fully installed casing. The drive pit must be sufficiently large to facilitate the ramming operation. It should be designed to include the rammer, support structure, welding pit, and working space for several workers. Personnel working in the pit include the operator of the pipe ramming machine plus two to three individuals, depending on the particular project, for such tasks as welding pipe sections and removing spoils. The pit should be properly shored to prevent collapsing. Fig. 5-1 illustrates a typical cross section of a launch pit.

The excavation of the pits should be accomplished using the proper technique and equipment, which depend on the size of the pits. The pits should not be excavated to be unnecessarily deep or larger than required to properly perform their functions. During and after the excavation, proper techniques should be used to protect the walls of the pit, such as steel soldier piles and horizontal wood lagging or sheet piling. For more information on pit excavation, refer to Smith and Andres (1993).

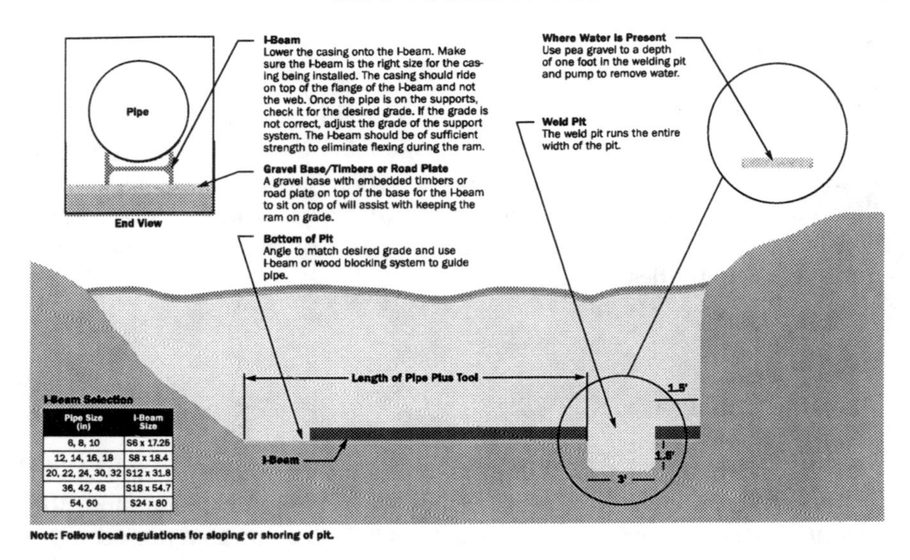

Pipe Size (in)	I-Beam Size
6, 8, 10	S6 x 17.25
12, 14, 16, 18	S8 x 18.4
20, 22, 24, 30, 32	S12 x 31.8
36, 42, 48	S18 x 54.7
54, 60	S24 x 80

FIGURE 5-1. Typical entrance pit before pipe insertion and equipment placement.

5.2 WORK PLAN

The work plan should include or address the following items:

- permits;
- excavation techniques;
- responsibilities for design adequacies;
- supervision;
- inspection;
- protection of adjacent structures;
- public safety;
- ventilation requirements; and
- method of spoil removal from pit (e.g., skid steer, backhoe, crane and bucket, vacuum, auger boring, or pigging.).

Contingency plans should be required in the event of unexpected subsurface conditions, obstructions, or contaminated ground conditions. The contractor must comply with all OSHA rules, as well as the guidelines provided in the ramming manufacturer's equipment safety and operations manual. As appropriate, all local, department of transportation, and railroad guidelines and requirements for depth of cover shall be followed during the installation.

5.3 EQUIPMENT

Equipment should be selected based on the requirements of the pipe ramming project, including the casing diameter, project length, required or anticipated air pressure and consumption, space restrictions, and other special design requirements or restrictions. The manufacturers' specifications provide equipment dimensions, power requirements, and equipment capabilities.

5.4 LUBRICATION SYSTEMS

The lubrication fluids used in pipe ramming operations are applied in the annular space between the casing pipe being installed and the surrounding soil. Their primary function is to reduce friction between the surface of the casing pipe and soil so as to reduce required push forces by the pipe ramming equipment. Lubrication may also be used to facilitate the installation of the product pipe inside the casing.

5.4.1 Lubricants

The lubrication fluid is used within the annular space during the ramming operation. The lubrication fluid is typically composed of bentonite, bentonite–polymer, or polymer mixtures. When bentonite is properly mixed, it is sheared apart into small, flat, thin platelets. These bentonite platelets act as small shingles and essentially plaster or seal the inside of the borehole by forming a wall or filter cake. This filter cake is of sufficiently low permeability to inhibit the fluid from permeating out of the formation, allowing it to remain within the annular space for an extended period. In sand, the bentonite platelets fill the pore spaces between the sand grains. This action allows the bentonite platelets to form a groutlike layer, greatly reducing the effective porosity and permeability of the sand. The filter cake has excellent lubrication qualities. For clay or shale, polymer-based lubricants would generally be used to avoid the tendency for swelling and tackiness that typically characterizes native clay and shale.

Because the water phase of the fluid is not entirely stopped from filtering into the surrounding formation, for larger diameter pipes the lubrication ports must continuously replenish the lubrication fluid that has filtered out over the longer time required to install the larger casings. The amount of lubrication fluid needed is a function of the soil type and volume of the annular space.

5.4.2 Lubrication Equipment

The lubricant is usually blended using a lubrication mixer, and it flows through ports mounted along the casing. The lubrication fluid is pumped to the exterior of the casing through openings located at regular intervals. Lubrication is recommended for essentially all installations, with the exception of very short or very small-diameter pipes. Figs. 5-2 and 5-3 present a typical lubrication mixer and lubrication pump, respectively. Fig. 4-1 in Chapter 4 illustrates a typical lubricating line installation.

FIGURE 5-2. Typical lubrication mixer.

FIGURE 5-3. Typical lubrication pump.

5.5 PRODUCTIVITY

The productivity of a pipe ramming operation depends on the site conditions, required accuracy of the pipeline grade, project-specific conditions, welding and set-up time, size and quality of the casing pipe, required pipe ramming equipment, support equipment, and the experience of the contractor and the operator. Typical productivity rates should exceed 4 ft (1.2 m) per hour. Corrective action should be considered if the production proceeds at significantly less than this rate. Possible remedial actions include using a larger capacity (energy or push force) rammer, periodically removing spoil from inside the pipe, or using improved lubrication.

5.6 DEWATERING

Dewatering should be performed in locations with a high groundwater table to prevent flooding and facilitate the pipe ramming operation. The contractor is generally responsible for the design, installation, operation, and maintenance of a dewatering system that meets the needs of the construction method. During the pipe ramming operation, the water level should be at least 2 ft (0.6 m) below the bottom of the casing or pit bottom, as appropriate.

Factors affecting the appropriate method of dewatering include the following:

- the nature of the soil and its hydrologic properties,
- the size and depth of the excavation,
- the proposed methods of excavation and ground support,
- the proximity of existing structures and their depth and type of foundation, and
- the nature of possible contamination at the site.

5.7 INSTALLATION OF CASING

The pipe ramming operation begins with the casing resting on its guidance rail. Construction personnel must not stand on the pipe during the installation. If necessary to provide initial resistance in difficult soil conditions, a bucket of an excavator may be used to apply lateral weight or pressure on the casing.

It is critical that the proper line and grade be maintained within the first 4 ft (1.2 m) of pipe insertion into the soil. If the grade is not correct after 4 ft (1.2 m) has been placed, it is highly unlikely that the proper line and grade can be achieved, even if the casing is pulled back in an attempt to correct the installation. In such an event, the face of the bore at the pit has been disturbed, and the loose soil at the face of the bore inhibits the ability of the operator to control and maintain the proper line and grade for the overall installation. The recommended procedure therefore includes maintaining close control of the air supply to the ramming tool at the start of the installation so that the pipe progresses slowly. The slow movement allows the operator to monitor line and grade at 1-ft intervals to achieve the required line and grade.

After the first section of casing has been inserted, the ramming mechanism is reversed and the pneumatic or hydraulic power supply is shut down, and an additional section of casing is placed in the guide rail and aligned with the first casing. The adjacent casings may be welded or mechanically connected by an interlocking press-fit (Permalok brand) connection. If the first section of casing has a lubrication line installed, the two lubrication lines are tacked together. When the new section of casing is properly placed and jointed, the pipe ramming operation resumes. The process is repeated until the casing installation is completed.

5.8 INSPECTION AND MONITORING

A knowledgeable qualified person should perform on-site inspections and monitoring of the operation, including a thorough inspection of the track position relative to the specified line and grade, before the casing enters the embankment. After the casing is installed and before the carrier pipe is installed (if required), the following steps should be performed:

- Proper position and usage of blocking and spacers to be used on carrier pipe should be verified, if necessary (see Section 3.3.1 of Chapter 3).
- The amount of spoil removed should be examined and quantified as a means of identifying possible voids external to the casing, although this is unlikely for a proper ramming operation. Any voids need to be grouted using approved grouting materials and methods.

5.9 AS-BUILT DRAWINGS AND DOCUMENTATION

The following information should be determined and documented by the contractor before performing the pipe ramming operation:

- length of the bore and the proposed line and grade, as determined by a detailed survey and
- the entry and exit points of the bore.

Following the pipe ramming operation, as-built drawings and information should be provided by the contractor documenting the installed position, including depth, line and grade of the casing, and internal carrier pipe, if appropriate. Any suspected damage to existing utilities must be indicated.

The contractor should provide a surface monitoring system, including the establishment of survey points on the surface to check possible settlement and heave. In general, local, state department of transportation, and railroad guidelines specify allowable tolerances. Other parameters that may be documented include the type and quantity of lubricant pumped and the quantity of spoil removed.

5.10 MEASUREMENT AND PAYMENT

The most common method of payment is based on a unit price per linear foot or meter of casing installed. Alternatively, payment may be made based on a previously agreed-on firm fixed price for which the scope of work has been clearly specified. In some cases, such as fast-track and emergency projects, cost-plus (time and material) may be appropriate. In most cases, the payment method and details are agreed on by the general contractor and the ramming subcontractor.

5.11 TYPICAL COSTS FOR PIPE RAMMING

Similar to productivity rates, project costs vary with soil conditions, project-specific conditions, and the overall scope of the job, including required accuracy of the line and grade and size and type of pipe casing. It should be noted that, for pipe ramming, contrary to most other projects, the price per foot typically increases as the overall length of the installation increases. This is because of the disproportionately greater risk associated with pipe ramming when attempting to achieve longer installations, including the need for more robust (stronger, with thicker walls) casings and proper pipe ramming tools, as well as support equipment and slower production rates.

5.12 SAFETY ISSUES

Safety on the jobsite is primarily the responsibility of the contractor (and subcontractor, as appropriate). The contractor must adhere to all applicable OSHA regulations, as well as guidelines provided in ramming manufacturer's safety and operations manuals.

CHAPTER 6
CASE STUDIES

6.1 CASE STUDY NO. 1: PIPE RAMMING HELPS RELIEVE FIRE AND FLOOD PROBLEMS IN LOS ALAMOS

The Cerro Grande fire ravaged the Los Alamos, New Mexico, landscape in May 2000. In addition to threatening the world-famous Los Alamos National Laboratory, the fire storm consumed more than 47,650 acres (19,283 hectares) of forest and left more than 400 families homeless. Almost as soon as the fire was contained, a new threat arose: flooding.

The massive fire left the mountainous region vulnerable to excessive water runoff. The threat of massive summer flooding prompted the Los Alamos National Laboratory and local communities to take steps to help mitigate the potential disaster. New drainage pipes had to be installed to help route the expected runoff.

The first project called for the installation of a shallow 250-ft (76-m), 36-in. (915-mm) steel drainage casing under a roadway (Fig. 6-1). The soil at the jobsite was a sediment-rich soil, which made ramming a challenging operation. Fig. 6-1 illustrates the 250-ft (76-m) casing installation, which took place under one of only two roads leading to the Los Alamos National Laboratory. Traffic disruption was avoided through the use of trenchless pipe ramming.

The contractor originally planned for the use of trenchless auger boring, but plans changed dramatically once different soil conditions were encountered. Therefore, the contractor decided on pneumatic pipe ramming.

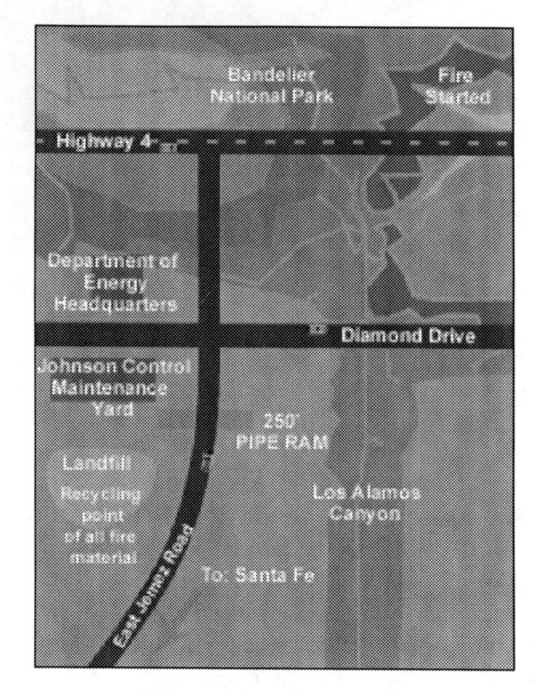

FIGURE 6-1. The 250-ft (76-m) casing installation.

6.1.1 Fire and Water

Ironically, the Cerro Grande fire was a result of a controlled burn designed to help prevent just such an incident. Although the laboratory basically escaped the fire, the area's watershed and natural erosion controls were not as lucky. The fire storm destroyed thousands of acres of forest. The underbrush in that forest slows water runoff.

In addition, the intense heat of the fire transformed the soil. The charred soil became unable to absorb water readily. For an area that averages more than 8.5 in. (216 mm) of rain from July through September, these conditions signaled potential disaster.

Efforts to reduce the impact of potential flooding began almost immediately. The restoration process began with cutting down charred trees to help divert runoff, tilling the fire-baked soil, and planting quick-growing cover vegetation, such as rye and barley, to soak up rainwater.

The laboratory took steps as well, identifying drainage routes and contracting to have drainage casings installed and replaced. After the fire, local representatives sought out roadways that experienced excessive water runoff and designated them for drainage casings. The job took place under a roadway just a few blocks from the laboratory and about a mile and a half from the fire itself. When installed, the new casing would channel the water under the road and on its way down the mountain.

6.1.2 Change of Plans

According to the contractor, the 250-ft (76-m) installation was originally intended to be an auger bore through volcanic ash-type soil. After digging a test hole, it became apparent that auger boring was not going to work. "We could barely dig through the ground with a 90,000-lb (40,823-kg) track-hoe. The soil was a densely compacted cobble conglomerate with large boulders, some as large as 24 in. (610 mm) in diameter. We found that out later."

The original project specification called for 250 ft (76 m) of 36-in. (915-mm) 0.375-in. (9.5-mm) wall casing. When it became apparent that augering wasn't going to work, a specialist from the supplier of the pipe ramming system provided technical support and put in a change order for a thicker walled casing to accommodate pipe ramming.

6.1.3 Ramming Basics and Benefits

Compared to other construction operations, pipe installation through ramming can be considered a simple method: A pneumatic hammer is attached to the rear of the casing or pipe. The ramming tool, which is basically an encased piston, drives the pipe through the ground with repeated percussive blows.

During the pipe ramming, a cutting shoe is often welded to the front of the lead casing to help reduce friction and cut through the soil. Bentonite or polymer lubrication can also be used to help reduce friction during ramming operations.

Several options are available for ramming various lengths of pipe. An entire length of pipe can be installed at once or, for longer runs, one section at a time can be installed. In that case, the ramming tool is removed after each section is in place, and a new section is welded onto the end of the newly installed section. The ramming tool is connected to the new section, and ramming continues. Depending on the size of the installation, spoil from inside the casing can be removed with compressed air, water, an augering system, or with a minibackhoe.

Some casing installation methods are impaired or even rendered inoperable by rock- or boulder-filled soils. Pipe ramming is different. During pipe ramming, boulders and rocks as large as the casing itself can be swallowed up as the casing moves through the soil and can be removed after the installation is complete.

Ramming tools are generally capable of installing 4-in. (100-mm) to 80-in. (2,032-mm) diameter pipe and steel casings. Diameters up to

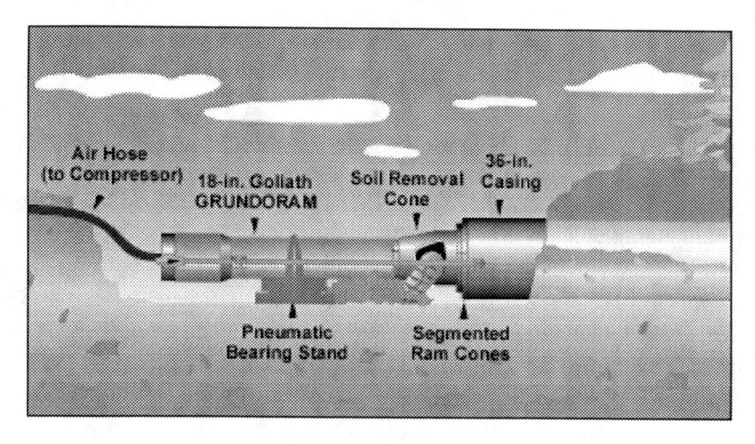

FIGURE 6-2. A typical pipe ramming setup.

148 in. (3,759 mm) have been successfully installed using large-scale ramming equipment. Ramming requires minimal working depths and is proven effective for horizontal, vertical, and angled applications. Ramming is also ideal for installations under roads and railroads because it displaces the soil without creating voids or slumps. Figure 6-2 presents a typical pipe ramming setup.

The conditions at Los Alamos put the pipe ramming method to the test. For the Los Alamos job, the construction crew was able to install 250 ft (76 m) of 36-in. (915-mm) casing through extremely challenging soil conditions using an 18-in. (460-mm) diameter pneumatic pipe ramming tool, as seen in Fig. 6-3.

6.1.4 Project Preparation

Over the 250-ft (76-m) run, the construction crew needed to maintain a 0.1% grade. The crew dug a 60-ft (18-m) long launch pit to accommodate the ramming equipment and the 40-ft (12-m) long sections of casing. "After the pit was dug, we moved the first casing section into position. We checked line and grade using surveyor's equipment and water levels." Once the casing was in position, the crew began assembling the ramming gear to connect the 18-in. (458-mm) diameter pipe ramming tool with the 36-in. (915-mm) diameter casing.

The pipe ramming tool selected weighs approximately 5,400 lb (2,450 kg) and operates at 1,236 ft^3/min. (35 m^3/min.). At full force, the piston moves at 180 strokes per minute. A pneumatically powered adjustable bearing stand was used to raise the tool to the required height for ramming.

FIGURE 6-3. Casing installation with pipe ramming.

To connect the ramming tool with the casing, a series of tapered and segmented cones are used. The configuration for the Los Alamos National Laboratory included a segmented ram cone, as well as a soil-removal cone. When assembled, the segmented ram cone reduces the overall diameter from 36 in. (915 mm) to approximately 24 in. (610 mm). The soil-removal cone is then added and further reduces the diameter to approximately 18 in. (458 mm). At this point, the tool is connected, friction fit, to the soil-removal cone, thus completing the assembly.

6.1.5 Pipe Ramming Operation

Once the connection between the tool and the casing was complete, the crew was ready to begin ramming. However, the crew encountered some giant boulders about 60 ft (18 m) into the run, which had to be removed before ramming could start. After these boulders were removed, the pipe ramming operation started and any remaining boulders were swallowed by the leading pipe.

The crew pumped approximately 100 gal. (378 L) of bentonite and water per 40-ft (12-m) casing section. Once a section was installed, welds between casings took anywhere from 3.5 to 4 hours to complete. Ramming times averaged 1 ft (0.3 m) every 7 min. in the beginning of the run and slowed to 1 ft (0.3 m) every 20 min. at the end of the run.

Once the 250-ft (76-m) casing was in place, the crew began removing the spoil with an auger system. "We went in there with a lead auger of 24 in. (610 mm) and removed as much as we could. After that, we went in with a 30-in. (762-mm) auger and removed additional spoil. Then we finished up with our 36-in. (915-mm) auger."

Overall, everyone was impressed by the equipment performance. The contractor could barely dig through the ground with a backhoe. The soil conditions were incredibly difficult. The contractor had expected more pipe ramming work in the Los Alamos area and at the time was planning for a 72-in. (1,830-mm) ram.

6.2 CASE STUDY NO. 2: A RECORD RAM

In 2004, a world pipe ramming record was achieved at a project in Altoona, Iowa. The contractor was required to install a 60-ft (18-m) tunnel a few feet beneath a rail line to facilitate a new bike path. The project ranks as the largest diameter pipe ram ever completed until that time. According to the contractor, the installation method was a key to the project's success. "During the early stages of this design, the engineering company contacted us regarding jacking in a square concrete tunnel underneath railroad tracks for a bike tunnel. We have completed both square and round projects in the past. After looking at the site, we determined that a square tunnel would not be a good choice for this project as the tracks were built on a levee. Our experience has been that levees are typically built on or with fill material that would hinder this type of trenchless application. We decided to pipe ram a round casing instead."

To create a tunnel large enough to accommodate bicyclists, the steel casing would need to be large. The 147-in. (3,734-mm) outside diameter casing ultimately used for the tunnel a new world standard for pipe ramming diameter, besting the previous record by 3 in. To complete the job, the contractor selected a 24-in. (610-mm) diameter pneumatic pipe rammer.

To facilitate the ramming of such an enormous casing, the contractor had an inverted bell ramming adapter (Fig. 6-4) fabricated by Arntzen Steel of Rockford, Illinois. The 147 in. (3,734 mm) reduced the overall diameter to 80 in. (2,032 mm). An 80-in. (2,032-mm) ram cone and a 24-in. (610-mm) ram cone were added to make the connection to the pipe ramming tool.

In preparation for ramming, the crew placed a 50 × 16 ft (15 × 5 m) concrete pad (Fig. 6-5). The pad served as the support for ramming operations, including a push sled and an auger track.

6.2.1 Ramming Basics

Ramming tools in general are capable of installing 4-in. (100-mm) through 122-in. (3,100-mm) diameter pipe and steel casings. At 24-in.

FIGURE 6-4. Bell ramming adapter.

FIGURE 6-5. Concrete pad.

(610-mm) diameter (Fig. 6-6), the selected tool for this project was the world's second-largest pipe rammer. The world's largest ramming tool is 32 in. (813 mm) in diameter. Before this project, casing diameters up to 144 in. (3,658 mm) had been successfully installed using large-scale ramming equipment.

FIGURE 6-6. A 24-in. (610-mm) diameter ramming tool.

6.2.2 Project Preparation

The ram took place under railroad (which remained active throughout the project) owned by the Iowa Interstate Railroad Ltd. A launch pit was dug on the south side of the tracks to accommodate a 50 × 16 ft (15 × 5 m) concrete pad that would serve as a platform for operations. The contractor said, "With an existing park on the north side of the track, we didn't want to cause any more environmental impact to the area than was necessary. We built a concrete backstop and concrete launch pit on the south side of the tracks, where excavation and clearing would take place anyway for the new bike trail."

After the pit construction was completed, the crew assembled a driving stage for the Taurus from an auger track. The crew also used a hydraulic push sled to assist with ramming operations. "We used the boring machine to push against the casing during ramming. We also used it to move the rammer back after each segment of pipe was rammed into place so the next piece of casing could be brought in."

The pipe chosen for the project was made by Permalok of St Louis, Missouri. The casing was fabricated in 20-ft (6-m) sections. Each 20-ft (6-m) section had a wall thickness of 1.5 in. (38 mm) and weighed 2,330 lb/ft (3,524 kg/m) or approximately 46,600 lb (21,317 kg) per section. The

Permalok brand casing incorporates a mechanical press-fit design without an internal or external bell, dramatically reducing the connection time otherwise required by eliminating welds and X-ray inspections. The crew carefully lowered a 20-ft (6-m) long 147-in. (3,734-mm) diameter pipe section into place, and after the prep work was complete, ramming was ready to begin.

6.2.3 Ramming Operation

The connection between the 24-in. (610-mm) diameter ram and the 147-in. (3,734-mm) outside diameter casing was made using a special adapter. The 147-in. (3,734-mm) inverted bell pipe adapter reduced the overall diameter to 80 in. (2,032 mm). An 80-in. (2,032-mm) ram cone was then connected to the adapter and further reduced the diameter to 30 in. (762 mm). A 24-in. (610-mm) ram cone made the final connection to the tool. The entire configuration was secured with tensioning chains, and the tool was connected to the air compressor.

The ramming went smoothly. Crews were able to install the first 20-ft (6-m) section of casing without incident. The ramming tool was removed, and some of the spoil was taken out of the casing. Crews then positioned the next 20-ft (6-m) section of pipe in place and made the connection to the first section. The rammer was reconnected, and the second section was rammed in place. The entire ram was completed on a 2.2% downhill grade. Once all of the casings were installed, the crew removed the remaining spoil with a skid steer loader. Almost 6,000 ft^3 (170 m^3) of soil was removed from the casings.

The 147-in. (3,734-mm) outside diameter casing (Figs. 6-7, 6-8, and 6-9) represents the world's largest diameter casing successfully rammed, besting the old mark by 3 in.

FIGURE 6-7. The 147-in. (3,734-mm) outside diameter casing.

FIGURE 6-8. This giant casing was rammed 60 ft (18 m) under Iowa Interstate Railroad tracks.

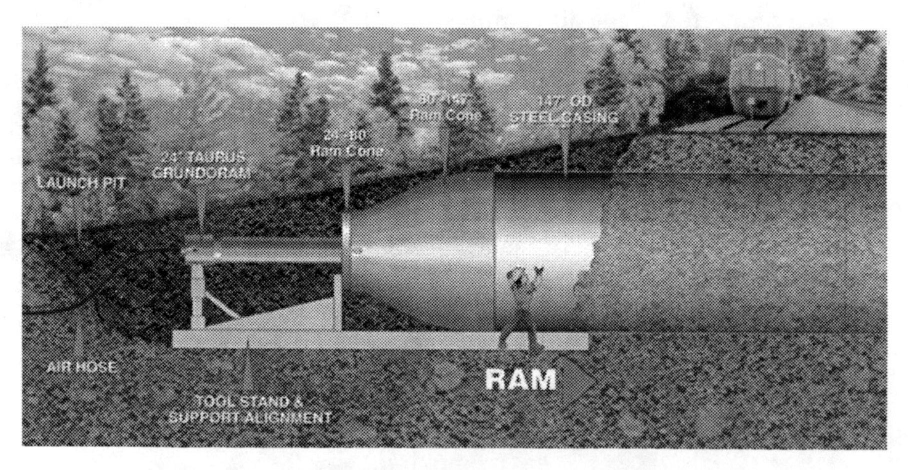

FIGURE 6-9. A ramming operation.

6.2.4 Project Completion

The contractor attributed the success of the project to several factors: "The willingness of the engineers to look beyond conventional construction techniques and allow a trenchless solution to be developed was the first part of the success. The crew was professional and courteous at all times. The railroad companies were cooperative and were watching this project very closely. They couldn't believe it when we rammed the pipe through to the exit side and never disturbed the tracks."

Safety is the most important factor. There were no accidents on this project. The crew even built a guardrail around the launch pit to ensure the safety of the many spectators this project had. Hard hats, safety vests, safety glasses, earplugs, and steel-toe boots were worn by all workers and even by spectators.

The giant casing was rammed 60 ft (18 m) under Iowa Interstate Railroad tracks. The tracks remained open during the ramming operations. The casing was installed to create a tunnel under the tracks for a new bike path for the city of Altoona, Iowa.

6.3 CASE STUDY NO. 3: GAS MAIN INSTALLATION PROVES THE POWER OF PIPE RAMMING

Railroad crossings pose a challenge for any utility contractor. Surface slump and track settling caused by a utility line installation can make a routine project extremely costly. A contractor in the state of Washington used pneumatic pipe ramming on a number of gas main installation projects for Puget Sound Energy (PSE), of Bellevue, Washington.

The contractor said that trenchless technology has allowed their crews to improve their efficiency and broaden their capabilities. "We have completed ramming projects in the past and in order to complete the installation of the high-pressure gas main for Puget Sound Energy, we needed to install a protective casing. We looked at several options, but ultimately pipe ramming was selected as the best option for installing the casing without disturbing the tracks or nearby utilities." For the ramming project in Snoqualmie, Washington, an 8.5-in. (216-mm) diameter pneumatic pipe ramming tool was used.

6.3.1 Contractor's Background

The contractor has been serving the needs of the telecommunications, gas, and oil industries throughout the western United States for more than 20 years and has worked closely with clients on a regular basis from design and material recommendations through the construction phase, all in an effort to minimize the costs of construction.

The firm used a wide range of construction techniques to complete projects. These techniques ranged from trenching in highly congested metropolitan areas to trenchless methods, such as directional drilling, pipe bursting, pipe ramming, and plowing cable in rural locations.

6.3.2 Snoqualmie Project Background

The contractor's first ramming project was designed to help meet the growth expectations for the city of North Bend, Washington. Puget Sound Energy wanted to extend its high-pressure gas line to the community of

North Bend. At the time, North Bend had an intermediate-pressure line serving the area. With fast growth, the gas line was undersized, and they also had cold-weather problems. The gas company crew had to provide compressed natural gas trailers to reinforce the gas service. So the objective of the project was to provide a high-pressure feed. Once in place, the town would have a 250-lb/in.^{2} (1,724-kPa) system rather than the undersized 55-lb/in.^{2} (379-kPa) system that was serving the area at the time.

To reach the community, a new utility line needed to cross under a set of railroad tracks located at the intersection of Snoqualmie Parkway and Highway 202. According to the contractor, the project location was also a tourist attraction. The Northwest Railway Museum operates a small rail line called the Snoqualmie Valley Railroad. The museum offered small excursions on the line to places like Snoqualmie Falls, another top tourist attraction. To facilitate the installation of the new 8-in. (200-mm) high-pressure steel gas main, several feet beneath the railroad tracks, the crew needed to first install a 120-ft (36-m) long, 12-in. (305-mm) steel casing to house the new line.

In addition to the tracks, the crews needed to be as precise as possible with the installation because of adjacent utilities. "We had to thread a needle. We had several drainage culverts above us, and an un-located water main in the vicinity. In addition, we were required by PSE to install the casing with a 2% downward grade to allow drainage."

6.3.3 Project Description

At the crossing location, the crew faced a series of obstacles. The crew began by excavating a launch pit on one side of the tracks. But being in the flood plain of the Snoqualmie River, the ground was saturated and needed to be dewatered. The crew installed two dewatering wells to stabilize the ground conditions of the flowing sand. Once the ground conditions were stabilized, the crew excavated the pit to a depth of approximately 15 ft (4.5 m) and placed shoring boxes. After the shoring was completed the crew used 4-in. (100-mm) and 6-in. (150-mm) riprap to create a base for the ramming tool.

The contractor said, "For the ramming portion, we set an I-beam down on the riprap base. The beam acted like a sled for the pipe. Because of the limited amount of space, we were only able to ram 20-ft (6-m) sections of pipe at a time."

After setting the first section of casing in the pit, the pipe rammer was brought in and connected to the casing through the soil port. To make the connection between the 8.5-in. (216-mm) diameter pipe rammer and the 12-in. (305-mm) diameter pipe, a soil port was attached, which made the transition between the tool and the pipe. The soil port also helped reduce the head pressure created by the spoil inside the pipe by giving it a place to escape. For larger casing diameters, a series of segmented ram cones and sometimes ramming adapters were used to make the connection between the ramming tool and the casing.

With the tool connected and in position, ramming proceeded without any problem. Once a 20-ft (6-m) section of casing was installed, the crew would remove the rammer and position the next 20-ft (6-m) section of the pipe. The new pipe section was then welded onto the end of the installed section. In this case, the crew performed complete penetration welds, and welding times ranged from 20 to 30 min. Once welding was complete, the rammer was reconnected with the pipe and ramming continued. Overall, the crew rammed six 20-ft (6-m) sections of casing at a 2% grade.

6.3.4 Spoil Removal

Cleaning out the spoil in the pipe became difficult. The contractor first tried to remove the spoil with vacuum track, but found that after the sand was dewatered, it was too hard to vacuum. So the contractor ended up bringing in a high-pressure water jet to clean out the pipe. The unit had jet nozzles with a 45-degree angle. Through the force of the water, the unit burrowed its way through the sand, and water jets forced the sand out of the pipe. Once the spoil was removed, the crew was able to install the new 8-in. (200-mm) gas main.

The contractor was pleased with the speed and success of the project. "We started ramming on a Monday afternoon. We were able to punch in a couple of joints. Then we came the next day and knocked the rest of them through. It was so fast, I had people that were interested in coming down and seeing the process and I had to call them and say, 'Sorry, it's done already.' "

6.3.5 Second Pipe Ramming Project

Pipe ramming was also the method of choice for another casing installation project performed by the same contractor, this time in Seattle. Whereas some casing installation methods are impaired or even rendered inoperable by rock or boulders, during pipe ramming, boulders and rocks as large as the casing itself can be swallowed up as the casing moves through the soil and can be removed after the installation is complete.

A cutting shoe is often welded to the front of the lead casing to help reduce friction and cut through the soil. Bentonite or polymer lubrication can also be used to help reduce friction during ramming operations.

While installing a 10-in. (254-mm) steel casing to house a 6-in. (150-mm) medium density polyethylene (MDPE) pipe gas main in Seattle, the crew encountered some difficult conditions. The project was underneath two of Amtrak's live rails and four future tracks. It was one of the older industrial areas, and when the crew excavated the pits, they discovered bricks and other debris. The contractor ended up ramming casing through a concrete wall and a mix of bricks and other construction materials. At the conculsion of ramming operation, they used a directional drill rig to remove the debris out of the pipe.

GLOSSARY

Advancement rate: Speed of advance of a pipe ramming or other trenchless construction through the ground, generally expressed as feet per day (or meters per day).

Annular space: The space surrounding one cylindrical object placed inside another.

Backstop: Also thrust block or dog plate; a reinforced area of the entrance pit wall directly behind the track or where the jacking loads will be resisted.

Band: Also shoe; a ring of steel welded at or near the front of the lead section of casing to cut relief and strengthen the casing (used in horizontal auger boring).

Bentonite: Colloidal clay sold under various trade names that forms a slick slurry or gel when water is added; also known as driller's mud (see Drilling fluid or mud).

Bore: A generally horizontal hole produced underground, primarily for the purpose of installing services.

Boring: (1) The dislodging or displacement of spoil by a tunnel-boring machine (TBM), rotating auger, or drill string to produce a hole or bore. (2) An earth-drilling process used for installing conduits or pipelines. (3) Obtaining soil samples for evaluation and testing.

Boring pit: Also drive pit, entry pit, launch pit, or thrust pit; an excavation in the earth of specified length, depth, and width for placing the boring machine on required line and grade.

Carrier pipe: Also product pipe; an internal pipe within the steel casing, providing the desired utility service.

Cased bore: A bore in which a pipe, usually a steel sleeve, is inserted simultaneously with the boring operation. Usually associated with horizontal auger boring or pipe ramming.

Casing: Also casing pipe; a pipe used to line boreholes through which the product pipes, also known as carrier pipes or ducts, are installed. The casing itself usually serves as the final product pipe.

Casing pipe method: Method in which a casing, generally steel, is rammed or jacked into place, within which a product or carrier pipe is later inserted.

Circumferential: The perimeter around the inner surface of a circular pipe cross section.

Closed face: The ability of a tunnel-boring machine to close or seal the facial opening of the machine to prevent, control, or slow the entering of soils into the machine. Also may be the bulk heading of a hand-dug tunnel to slow or stop the inflow of material.

Cohesive soil: A clayey soil that when dried has considerable unconfined strength and acts as slurry when saturated.

Conduit: A broad term that can include pipe, casing, tunnels, ducts, or channels. The term is too broad to be used as a technical term in boring or tunneling.

Cradle: A structure constructed from concrete or masonry that provides structural support to a pipe. It typically surrounds the bottom and the sides of a pipe up to the springing line.

Cradle machine: A boring machine typically carried by another machine that uses winches to advance the casing.

Crossing: Pipeline installation in which the primary purpose is to provide one or more passages beneath a surface obstruction.

Dead man: A fixed anchor point used in advancing a saddle or cradle machine.

Directional drilling: A steerable system for the installation of pipes, conduits, and cables in a shallow arc using a surface-launched drilling rig. Traditionally, the term applies to large-scale crossings in which a fluid-filled pilot bore is drilled using a fluid-driven motor at the end of a bent sub; the hole is then enlarged by back reamer to the size required for the product pipe. The positioning of a bent sub provides the required deviation during pilot boring. Tracking of the drill string is achieved by the use of a downhole survey tool.

Dog plate: See Backstop.

Dogs: Movable protrusions in the backstop that engage holes or blocks in the track.

Drilling fluid or mud: A mixture of water and usually bentonite or polymer continuously pumped to the cutting head to facilitate cutting, reduce required torque, facilitate the removal of cuttings, stabilize the borehole, cool the head, and lubricate the installation of the product pipe. In suitable soil conditions, water alone may be used.

Drivepit: Also boring pit, entry pit, launch pit, or thrust pit; excavation from which trenchless technology equipment is launched for the installation of a pipeline, conduit, or cable. It may incorporate a thrust wall to spread reaction loads to the soil.

Dry bore: Any drilling or rod pushing system that does not use drilling fluid in the process. Usually associated with guided impact moling, but also some rotary methods.

Duct: (1) In many instances, a term interchangeable with pipe. (2) In the boring industry, it is usually used for small plastic or steel pipes that enclose wires or cables for electrical or communications uses. (3) Conduit inside which a utility service is installed.

Earth piercing: (1) Term commonly used in trenchless technology industry as an alternative to impact moling. (2) The use of a tool that includes a percussive hammer within a suitable casing, generally of torpedo shape. The hammer may be pneumatic or hydraulic. The term is usually associated with nonsteered devices without rigid attachment to the launch pit, relying on the resistance (friction) of the ground for forward movement. During operation, the soil is displaced, not removed. An unsupported bore may be formed in suitable ground, or a pipe is drawn in, or pushed in, behind the tool. Cables may also be drawn in.

Entrance pit: See boring pit, drive pit, entry pit, launch pit, or thrust pit.

Entry pit: See boring pit, drive pit, entrance pit, launch pit, or thrust pit.

Exit pit: See Exit shaft, Receiving pit, or Reception pit.

Exit shaft: See Exit pit, Receiving pit, or Reception pit.

Face: Wall of the entrance pit into which the bore is made.

Free boring: Also Uncased boring; a horizontal auger boring method without the use of casing installed at the same time as the hole is cut. This method should be used with caution because it is appropriate for only certain project and site conditions.

Guide rail: A device used to support or guide, first the shield and then the pipe within the drive pit during a pipe jacking or utility tunneling operation.

Guided impact moling: Guided impact moling uses walkover tracking and remote steering similar to that used in the horizontal directional drilling industry. A sonde integrated within the forward end of the tool body is ruggedized to withstand the hammering impact of the mole. The head of the mole is immovable and as such has a continuous tendency to steer. In order to bore approximately straight, the mole has to be rotated as it advances, and as a result a zigzag path is created when boring a nominally straight path.

Heaving: A reaction in which the ground in front of a tunneling or pipe jacking operation may be displaced forward and upward, causing an uplifting of the ground surface.

Height of cover: Distance from crown of a pipe or conduit to the finished road surface, or ground surface, or the base of the railway.

Impact moling: See Earth piercing.

Impact ramming: See Pipe ramming.

Jacking: The actual pushing of pipe or casing in an excavated hole. This is usually done with hydraulic cylinders (jacks), but it has also been accomplished with mechanical jacks and air jacks.

Jacking pipes: Pipe sections with smooth outside joints designed to be installed using pipe jacking techniques.

Jacking pit: See Jacking shaft.

Jacking shaft: A vertical excavation (such as a rectangular pit where length is slightly more than a pipe section length and depth is slightly more than pipe invert) from which trenchless technology equipment is launched for the installation or renewal of a pipeline.

Launch pit: Also known as boring pit, drive pit, or entry pit, but more commonly associated with launching a ramming or impact moling tool. The pit used for launching a trenchless technology operation. The launch pit is the start pit from which the pipe is being pushed or rammed.

Lubrication: The addition of water with a substance, typically bentonite or polymer, to reduce friction.

Man-entry: Also worker-entry; describes any inspection, construction, renewal, or repair process, which requires an operator to enter a pipe, ducts, or bore. OSHA currently has no minimum size limit for worker-entry operations; however, a much broader concept of confined space is addressed in Title 29 *Code of Federal Regulations*, Part 1910.146. The minimum size for which this is currently permissible in the United Kingdom is 36 in. (approximately 900 mm). Many trenchless technologies, such as pipe ramming method generally, do not require worker entry inside the pipe.

Midweld: Two or more pipe section joints welded to form a longer pipe section.

Moling: See Earth piercing.

No-dig technology: See Trenchless technology.

Obstruction: Any natural or synthetic object that lies on the path of a bore or tunnel and has the potential to stop the boring or tunneling operation.

Open-cut (conventional trenching): The conventional method by which access is gained to the required level underground for the installation, repair, or replacement of a pipe, conduit, or cable by excavation. The excavation is then backfilled and the surface restored.

Pipe invert: Lowest part of the inside wall of the pipe along its total length (opposite to crown).

Pipe lubricant: See Lubrication.

Pipe pusher or puller: A machine that pushes or pulls a rod or pipe to produce a bore by means of compaction without rotation or impact.

Pipe ramming: Also Impact ramming; a nonsteerable system of forming a bore by driving a closed-face or open-ended steel casing using a percussive hammer from a launch pit. The soil may be removed from the open casing by mechanical means, augering, jetting, or compressed air.

Product pipe: Also carrier pipe; permanent pipeline for operational use, a utility pipe for conveyance of water, gas, sewage, and other products.

Ramming: A percussive hammer is attached to an open-end casing, which is driven through the ground. See Pipe ramming.

Receiving: See Reception pit.

Receiving pit: (1) An opening in the earth located at the expected exit of the cutting head or tunnel-boring machine. (2) The pit that is dug at the end of the bore, opposite the jacking pit. (3) See Exit pit.

Reception pit: Excavation into which trenchless technology equipment is driven and recovered following the installation of the casing, product pipe, conduit, or cable. See Receiving pit.

Shoe: See Band.

Skin friction: Resistance to advancement caused by soil pressure and associated drag around the pipe or casing.

Sleeve pipe: A pipe installed as external protection for a product pipe.

Social costs: Costs incurred by society as a result of underground pipeline construction and renewal. These include, but are not limited to, traffic disruptions, environmental damage, safety hazards, inconvenience to the general public, and business losses resulting from road closures.

Spoil: Earth, rock, and other materials displaced by a tunnel, pipe, or casing and removed as the tunnel, pipe, or casing is installed.

Thrust: Force applied to a pipeline or drill string to propel it through the ground.

Thrust block: See Backstop.

Thrust pit: See Drive pit.

Trenching: See Open-cut (conventional trenching).

Trenchless technology: Also "no-dig"; the variety of techniques for underground pipeline and utility construction and replacement, rehabilitation, renovation (collectively called renewal); repair; inspection; leak detection; etc., with a minimum of excavation at the ground surface.

Uncased bore: A bore without a lining or pipe inserted; i.e., a self-supporting hole, either temporary or permanent. Typically not recommended, except under special circumstances. See also Free boring.

Underground utility: Active or inactive services or utilities below ground level.

Voids: (1) Holes external to the pipe in the surrounding soil or material. (2) A term generally applied to paints to describe holidays, holes, and skips in the film. (3) Shrinkage in castings or welds.

Worker-entry: See Man-entry.

REFERENCES

American Society of Civil Engineers. (1996). *Pipeline Crossings*, ASCE Manuals and Reports on Engineering Practice No. 89. ASCE, New York.

American Society of Civil Engineers. (2002). *Standard Guidelines for the Collection and Depiction of Existing Subsurface Utility Data*, CI/ASCE 38-02. ASCE, Reston, Va.

American Society of Civil Engineers. (2004). *Horizontal Auger Boring Projects*, ASCE Manuals and Reports on Engineering Practice No. 106. ASCE, Reston, Va.

American Society of Civil Engineers. (2007). *Geotechnical Baseline Reports for Construction—Suggested Guidelines*. ASCE, Reston, Va.

Moser, Al. (2001). *Buried Pipe Design*, McGraw-Hill, New York.

Najafi, M. (2005). *Trenchless Technology: Pipeline and Utility Design, Construction and Renewal*, McGraw-Hill, New York.

Smith, R. C., and Andres, C. K. (1993). *Principles and Practices of Heavy Construction*, 4th ed., Prentice Hall, Englewood Cliffs, New Jersey.

Watkins, Reynold, and Anderson, Loren. (2000). *Structural Mechanics of Buried Pipes*, CRC Press, Boca Raton, Fla.

INDEX